SELF-PUBLISH YOUR GENRE NOVEL

BEVERLEY EIKLI

Why I Wrote This Book

I've worked and lived all over the world, but my longest journey was getting my first novel published.

That journey took me twenty-seven years—and nearly a decade to realise what I'd been doing wrong…

I'd drowned the heroine on the last page.

So, what's wrong with that? In some of the greatest novels, a tragedy occurs on the final page.

In "Romeo and Juliet" by William Shakespeare, the star-crossed lovers take their own lives in a tragic misunderstanding, discovered in the final scene of the play.

In "The Great Gatsby" by F. Scott Fitzgerald, the main character, Jay Gatsby, is shot dead by George Wilson, who then takes his own life just before the novel's conclusion.

Why, then, was it so egregious that I drowned poor Miss Sophie Fortune on the last page of the first novel I shopped to at least fifty editors?

Finally, one kind editor sent me more than just a badly photocopied standard rejection and told me.

I was writing a romance, and the genre expectations for romance is a happy ending.

So, what about "Romeo and Juliet"?

That's a love story, not a romance.

These distinctions matter, which is why I'm writing this book.

I also want to share some of the pitfalls I encountered before finally having my first book accepted in a three-book deal, which led to five traditional publishing deals and a total of forty-one books published in three genres, both traditionally and self-published (or indie-published).

Doing both makes me what's called a hybrid author. While the traditionally published books have been fun experiences and given me the thrill of seeing my own books in print in bookshops and libraries, they haven't earned me the same royalties or control my indie-published books have.

In the following chapters, I'll share my experiences in traditional and self-publishing, explaining why I made certain decisions and the varying results. I'll also detail the expectations of genre fiction and outline some of the writing methods I've been teaching my Creative Writing students for the past ten years.

I'll talk about tropes, genre, categories and themes.

Finally, I'll explain the step-by-step process to get your book self-published and into the world, if you choose to take that path.

But whatever path you take, I have one caveat: Love what you do!

Writing can be hard, lonely work, though there can be amazing rewards that go beyond royalties. Friendship with other authors—people who "get" you—is one of these.

At the end of the day, however, you'll be putting in a lot of hours for perhaps minimal return unless you are committed to your passion and have a strategy.

So, take what inspiration you can as you read this "how-to," and follow me into a world of unknown possibilities...

Why Do You Write? And Are you a 'Plotter' or a 'Discovery Writer'?

Writing a genre novel is a significant undertaking that requires dedication and effort.

You'll have to set aside time from the other things you enjoy, and you'll need discipline.

It might even help to understand your motivations and what drives you to write.

Perhaps you want to share a message that's important to you. Maybe you are a more experienced writer whose success depends upon the many ways you craft your work to provide an escape for readers…

Or you might simply want to express your creativity and love for storytelling with no thought beforehand…Just the thrill of discovering where it takes you.

There's nothing wrong with that.

But if identifying your purpose for writing helps you stay focused and motivated throughout the writing process, then take a moment to reflect on what that is.

When you have a clear sense of why you're telling this story, it can become easier to make decisions about plot, character development, and themes. Your motivation can also guide you in establishing—beforehand—your target audience and tailoring your writing to their expectations and interests.

You're going to have to do that at some stage so beforehand makes sense.

Those are the hallmarks of a plotter who needs to know their story in advance.

A discovery writer (or a "pantser" as they are also known, since they 'write by the seat of their pants') writes for the joy of the journey. They let their characters lead the way.

Therefore a discovery writer will have to revise and edit more when they've finished their first draft which is when they identify theme and audience and make the according tweaks or edits.

I'd written many books before I realised that, although the plots were obviously different, the theme in each book was the same: redemption.

That obviously came with some juggling. A heroine who needs to be redeemed will have some flaws at the beginning while the purpose of the story is to see her grow, thus realising her deficiencies—and making positive change.

This was what tripped me up when I first started submitting my manuscript. My heroine was *too* flawed and therefore too unlikeable at the beginning. It turned readers—or, in my cases, judges—off before they got into the story.

Unless you can get your reader to identify with the protagonist's goals, or to accept that the reason for the book is to see them change, you will lose your audience.

So, don't make the mistakes I did.

Stay with me as we take a look at not just tropes, genre and theme, but of the expectations of the readers for whom those tropes, genres, and themes are so beloved.

Just as every writer is different, so is every reader.

But one thing I have learned as a genre writer is that to mess with reader's expectations when handling their favourite tropes is done **at your peril!**

Knowing Your Audience: Identifying and Understanding Your Target Readers

Who are the readers you hope to reach with your novel? Understanding your audience can help you make informed decisions about your story's content, tone, and style.

Start by researching the genre or subgenre you're writing in. Read widely within that category to familiarise yourself with the conventions, tropes, and reader expectations.

Consider the demographics of your target audience, such as age, gender, and reading preferences. This information can help you create characters and storylines that resonate with your intended readership.

It's also important to think about what your target audience wants from a story in your chosen genre.

Are they looking for fast-paced action, complex world-building, or emotional depth? Do they expect certain character archetypes or plot twists?

By understanding what your readers crave, you can craft a novel that satisfies their desires while still bringing your unique voice and perspective to the story.

So let's start with genre. What, exactly, *is* genre?

What is Genre?

The term 'genre' refers to a category or classification of artistic works, including literature, music, film, and other forms of creative expression. It categorises and organises works based on shared characteristics, themes, styles, or conventions.

In literature, genres include fiction, non-fiction, romance, mystery, science fiction, fantasy, and many others.

Each genre has its own set of conventions, tropes, and expectations. These conventions help readers or viewers understand what to expect from a particular work and allow for easier categorisation and identification.

Genres can overlap or blend together, giving rise to subgenres or hybrid genres. For instance, a book combining elements of mystery and fantasy might be categorised as a 'fantasy mystery'- or 'mystery with supernatural elements'.

Trends affect terminology, too. For example, a writer of novels combining fantasy and romance should research the search terms readers use to find their next binge-read.

In April 2024, statistics show that the search term "fantasy romance" clearly beats "romantic fantasy". However, in April 2023, the term "romantasy" took off and now accounts for 60% of searches in this genre (as of April 2024).

Later in the book, I'll explain why search terms are crucial for readers to find your book, leading to greater discoverability and sales. I'll also share different ways—free and paid—to research these search terms, also known as "keywords". When uploading your final edited genre novel to retailers, they will require you to supply relevant keywords for discoverability.

But back to Genre.

Genres serve several purposes. They provide a framework for artists to create within established boundaries or to innovate and challenge those boundaries. They help audiences find works that align with their preferences and interests.

Genres also facilitate critical analysis, discussion, and comparison among works with similar characteristics. This allows for a deeper understanding of artistic trends and themes within specific genres.

Genres evolve, and the popularity of certain genres waxes and wanes. However, understanding the genre you are writing, or aiming to write, is critical to publishing success.

Tropes, Genres & Themes

The terms 'trope' and 'genre' are related but distinct concepts within storytelling. Theme, is different, again. Here's an explanation of each:

Genre:

Genre refers to a broader category or classification of artistic works based on shared characteristics, themes, styles, or conventions. It provides a way to categorize and organize creative works into distinct categories.

Genres can encompass various elements such as setting, tone, narrative structure, and subject matter. Examples of genres include romance, science fiction, mystery, fantasy, historical fiction, cozy mystery, Gothic romance, and many more. Genres provide a framework for understanding and identifying works that share similar traits and allow audiences to find works that align with their preferences.

Trope:

A trope, on the other hand, refers to a recurring theme, plot device, character archetype, or narrative convention that appears across different works within a specific genre or across genres. Tropes are storytelling devices or techniques that convey familiar ideas, concepts, or patterns. They can be seen as the building blocks or tools that writers use to construct narratives. Tropes can be found within genres and can help define and characterise specific genres.

For example, the 'chosen one'- trope is often associated with fantasy stories, while the 'hard-boiled detective'- trope is commonly seen in mystery or noir genres. Tropes can be used in a variety of ways, from subverting expectations to reinforcing genre conventions.

Theme:

Theme is another important element in storytelling that relates to both genre and trope but has its own distinct role.

Theme refers to the underlying message, central idea, or moral lesson explored in a work of art, including literature, film, music, or visual arts. It is the deeper meaning or concept that the creator intends to convey through their narrative. Themes can be universal and transcend specific genres or tropes, as they often speak to broader human experiences, emotions, or social issues.

While genre provides the framework and categorisation of a work, and tropes offer recurring storytelling devices or conventions, themes delve into the deeper layers of meaning and exploration within the narrative. Themes can be present in a single work or can be explored across multiple works within a genre or even across various genres.

Themes can be explicit, where they are directly stated or explored in the narrative, or they can be implicit, where the reader or audience infers the underlying message or concept. They can address a wide range of topics, such as love, friendship, power, identity, justice, redemption, coming of age, or the human condition.

Themes can be used to enhance and enrich the storytelling experience. They provide depth, resonance, and emotional impact, allowing audiences to connect with the work on a deeper level and contemplate broader ideas or issues beyond the surface-level plot or genre conventions. Themes can be a driving force behind the narrative choices, character development, and overall meaning of the work.

So, while genre categorises works and tropes provide recurring devices or conventions, themes delve into the underlying messages, ideas, or moral lessons explored within the narrative. Themes can transcend genres and tropes, offering deeper insights into universal human experiences or social issues.

Structuring your Genre Novel

When it comes to crafting a compelling genre novel, the structure you choose can be just as important as the characters, plot, and setting. The way you organise your story can greatly impact its pacing, tension, and overall effectiveness in engaging your readers. In this chapter, we look at structures used in popular novels, while in the next, we'll discuss how these structures, such as the classic three-act structure to more complex non-linear narratives, work.

NOW, here's what's involved in each of these structures.

BELOW IS a graph of the Three-Act Structure

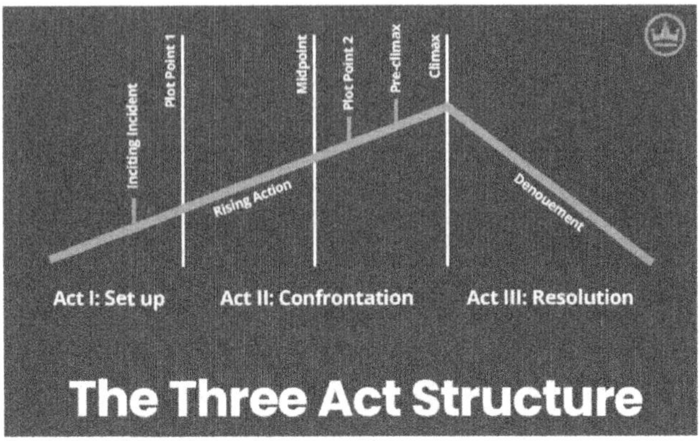

More details can be found here: https://kindlepreneur.com/three-act-structure/

And just as a side note, in addition to a huge archive or self-publishing advice, Dave Chesson is also the creator of Publisher Rocket, which I find indispensable software for researching Keywords which I use when uploading my novels to the retailers. More later in the book in the chapter on Useful Software for Authors.

THREE-ACT STRUCTURE

The three-act structure is a widely used approach in storytelling. It divides the story into three main sections: setup, confrontation, and resolution.

Act 1 (Setup):

Introduce the main characters, the world, and the central conflict. Set up the story's premise, establish the protagonist's goals, and create a hook to engage readers.

Act 2 (Confrontation):

Develop the conflict, raise the stakes, and introduce obstacles and complications. The protagonist faces challenges, undergoes growth, and experiences setbacks and conflicts.

Act 3 (Resolution):

Build towards the climax and provide a satisfying resolution. Resolve the central conflict, tie up loose ends, and provide closure for the characters and story.

THE HERO'S JOURNEY

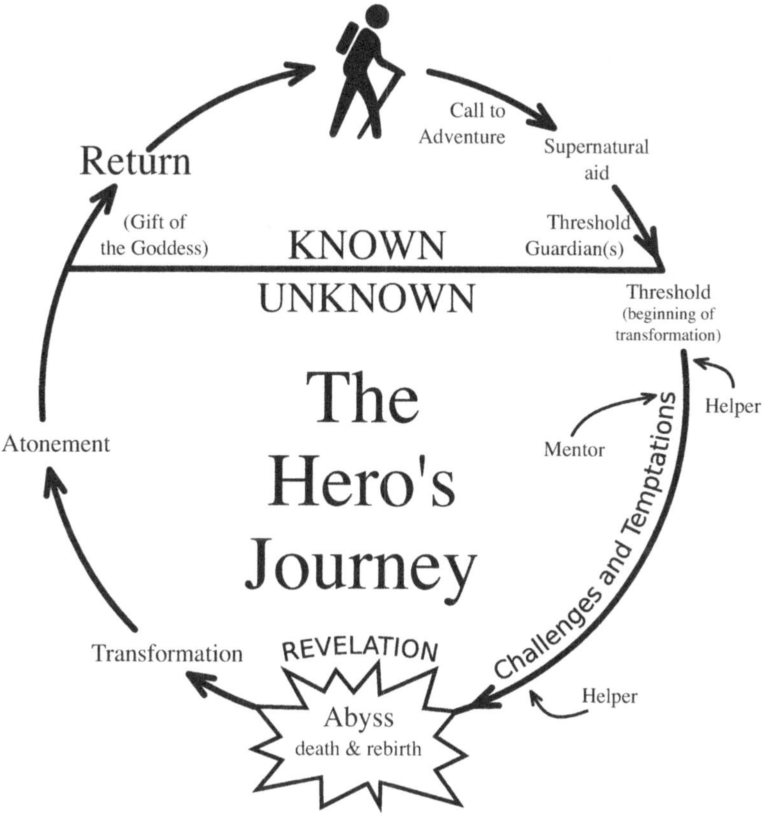

The Hero's Journey, also known as the Monomyth, is a common narrative structure identified by Joseph Campbell in his book "The Hero with a Thousand Faces."

It outlines the common stages and character development found in many stories across different cultures and time periods, and is a fundamental storytelling pattern that has been observed in ancient myths, folklore, and religious texts, as well as in modern literature and film.

It outlines the typical adventure of a hero who goes on a journey, faces challenges, and ultimately returns home transformed. Here are the key elements of the Hero's Journey:

1. Ordinary World: The hero's normal life before the adventure begins.

2. Call to Adventure: The hero is faced with a problem, challenge, or adventure that they must undertake.

3. Refusal of the Call: The hero initially hesitates or refuses the call to adventure due to fear, insecurity, or a sense of duty.

4. Meeting with the Mentor: The hero meets a mentor figure who provides guidance, training, and often gives the hero a gift or tool for the journey.

5. Crossing the First Threshold: The hero commits to the adventure and enters the special world, leaving their ordinary life behind.

6. Tests, Allies, and Enemies: The hero faces various challenges, makes allies, and confronts enemies while adapting to the special world.

7. Approach to the Inmost Cave: The hero approaches the central challenge or conflict of the adventure.

8. The Ordeal: The hero faces the greatest challenge yet and experiences a "death" (either literal or metaphorical), which leads to a transformation.

9. Reward (Seizing the Sword): The hero survives, overcomes the challenge, and often takes possession of a treasure or knowledge.

10. The Road Back: The hero decides to return to the ordinary world with their newfound wisdom or treasure.

11. Resurrection: The hero faces a final test where everything is at stake, using the lessons and skills gained during the journey.

12. Return with the Elixir: The hero returns to the ordinary world, but they are transformed by the experience and often bring something to share with others (a physical item or wisdom).

While not every story follows this structure exactly, understanding the Hero's Journey can help writers craft compelling character arcs and narratives that resonate with readers on a deep, universal level.

THE VILLAIN'S JOURNEY

This is also known as the "**Anti-Hero's Journey**" or "**The Tragic Hero's Journey**."

While the **Hero's Journey** follows a protagonist's path to growth, self-discovery, and ultimate triumph, the **Villain's Journey** tracks a character's descent into darkness, corruption, or tragedy.

Key elements of the **Villain's Journey** include:

- The character starts off as an ordinary or even sympathetic figure, rather than a predestined hero.
- The character is faced with challenges, temptations, or traumas that lead them down a darker path.
- The character succumbs to their flaws, makes morally questionable choices, or becomes corrupted by power or external influences.
- The character may experience a brief period of success or triumph in their villainous role.
- The character ultimately faces consequences for their actions, which may include downfall, punishment, or a tragic end.

EXAMPLES of the **Villain's Journey** can be found in characters like Michael Corleone in "The Godfather" trilogy, Walter White in "Breaking Bad," and Anakin Skywalker/Darth Vader in the "Star Wars" prequel trilogy.

In each case, the character's journey is marked by a transformation from a relatively ordinary or sympathetic figure to a villain, often through a series of moral compromises and damaging choices.

. . .

Nonlinear Structure:

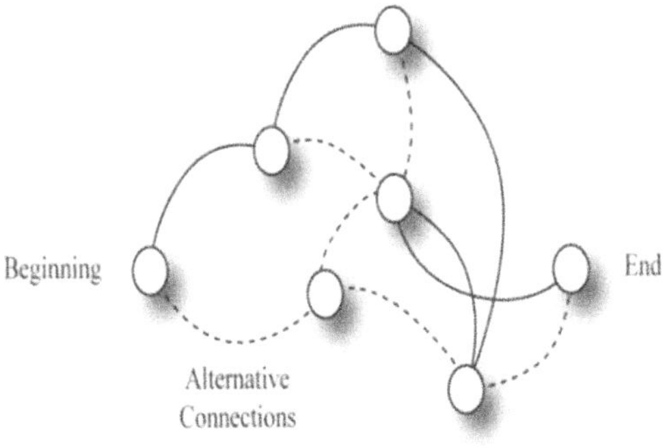

A nonlinear structure plays with the chronological order of events, using flashbacks, multiple timelines, or parallel narratives.

This approach can create suspense, intrigue, and complexity within the story. It allows for revealing information gradually or exploring different perspectives on the events.

Nonlinear structures are often found in mystery, thriller, and science fiction genres.

Episodic Structure:

An episodic structure divides the story into self-contained episodes or chapters, each with its own mini-story or conflict.

This approach is common in genres such as urban fantasy, detective fiction, or adventure series where the protagonist faces a series of episodic challenges or cases.

Each episode contributes to the overall story arc while providing a sense of progression and closure within individual segments.

MULTIPLE POV STRUCTURE

In a multiple point-of-view (POV) structure, the story unfolds through the perspectives of multiple main characters.

Each character's viewpoint offers unique insights into the plot, conflicts, and relationships. It allows readers to experience the story through different lenses and provides a broader understanding of the narrative.

This structure is often used in genres such as epic fantasy, mystery, or ensemble-cast novels.

These are just a few approaches to structuring a genre novel.
The choice of structure should align with the genre, story, and desired narrative effect. Feel free to adapt and combine these approaches to suit your specific needs and create a compelling and well-organised story.

The Epistolary Novel

An epistolary novel is a story told through a series of documents, usually letters, but can also include diary entries, newspaper clippings, emails, or other forms of communication. Here are the key elements required in an epistolary novel:

Multiple Correspondents: The story should be told through the writings of two or more characters, each with their own distinct voice and perspective.

Chronological or Thematic Order: The documents should be arranged in a way that allows the story to unfold naturally, either chronologically or thematically.

Character Development: The letters or documents should reveal the characters' thoughts, feelings, and motivations, allowing the reader to understand and empathize with them.

Plot Progression: The exchanges between characters should advance the plot, revealing key events, conflicts, and resolutions.

Absence of a Traditional Narrator: In an epistolary novel, there is no omniscient narrator guiding the story. The reader must piece together the narrative through the various documents.

Realistic and Distinct Voices: Each character's writing style should be consistent and reflect their personality, background, and relationship to the other correspondents.

Incomplete Information: Because the story is told through documents, there may be gaps in the narrative that the reader must fill in, creating a sense of mystery or ambiguity.

Intimacy and Immediacy: The epistolary format can create a sense of intimacy between the characters and the reader, as well as a feeling of immediacy, as if the events are unfolding in real-time.

Thematic Relevance: The choice to use an epistolary format should be thematically relevant to the story, enhancing the narrative's impact and meaning.

Climax and Resolution: Despite the unconventional format, an epistolary novel should still build toward a climax and provide a satisfying resolution for the reader.

Structures Used in Popular Books

The Narrative Structures used by the Bronte Sisters

Here's an explanation of the structures used in "Jane Eyre," "Wuthering Heights," and "Agnes Grey":

"Jane Eyre" by Charlotte Brontë - Linear Narrative with Bildungsroman Elements.

The story follows a linear narrative, tracing Jane's life from childhood to adulthood.

It is a Bildungsroman, a novel that focuses on the psychological and moral growth of the protagonist from youth to adulthood.

The novel is divided into three main sections: Jane's childhood at Gateshead and Lowood, her time at Thornfield as a governess, and her experiences after leaving Thornfield.

The story is told from Jane's perspective, allowing readers to closely follow her emotional and moral development.

"Wuthering Heights" by Emily Brontë - Frame Narrative and Non-Linear Structure.

The story is told through a frame narrative, with Lockwood, the tenant of Thrushcross Grange, learning about the history of Wuthering Heights and its inhabitants from Nelly Dean, the housekeeper.

The main narrative, as recounted by Nelly, follows a non-linear structure, moving back and forth in time to cover the lives of the two main families: the Earnshaws and the Lintons.

The novel spans two generations, focusing on the relationship between Heathcliff and Catherine Earnshaw, and later, their children.

The complex structure and multiple narrators create a sense of mystery and unreliability, mirroring the tumultuous nature of the characters' relationships.

"Agnes Grey" by Anne Brontë - Linear Narrative and Governess Novel

The story follows a linear narrative, chronicling Agnes Grey's experiences as a governess

It is a governess novel, a popular genre in the 19th century that explores the lives and challenges of young women working as governesses

The novel is divided into three main sections: Agnes's time with the Bloomfield family, her experiences with the Murray family, and her eventual resolution and personal growth

The story is narrated by Agnes herself, offering readers a direct insight into her thoughts, feelings, and moral development throughout her journey

All three novels use different structural elements to explore their protagonists' emotional and moral journeys. While "Jane Eyre" and "Agnes Grey" follow more linear narratives focusing on the growth of their title characters, "Wuthering Heights" employs a complex frame narrative and non-linear structure to create a sense of mystery and unreliability surrounding the tempestuous relationships at its core.

And three more because they are favourites of mine!

Examples of Different Structures in well known works

"The Girl on the Train" by Paula Hawkins

Multiple Perspectives and Non-Linear Narrative

The story is told from the perspectives of three women: Rachel, Anna, and Megan

Rachel's narrative is the primary focus, with her point of view being unreliable due to her alcoholism and memory lapses.

The timeline shifts between the present and the past, gradually revealing the events leading up to Megan's disappearance.

The non-linear structure and multiple perspectives create suspense and keep readers guessing until the mystery is resolved.

"GONE GIRL" **by Gillian Flynn**

Dual Narrative and Non-Linear Structure

The story alternates between the perspectives of Nick and Amy, the married couple at the center of the mystery.

In Part One, the narrative switches between Nick's present-day account of Amy's disappearance and Amy's past diary entries leading up to the disappearance

Part Two features a significant twist that changes the reader's understanding of the events and characters.

The non-linear structure and unreliable narration create a sense of uncertainty and suspense, as readers question the truth behind Nick and Amy's accounts.

The resolution in Part Three brings the two narratives together, revealing the complex and disturbing nature of their relationship.

BOTH "THE GIRL ON THE TRAIN" and "Gone Girl" use multiple perspectives and non-linear narratives to create psychological suspense and keep readers engaged. The unreliable narrators and shifting timelines contribute to the complexity of the stories, challenging readers to piece together the truth from the fragmented accounts presented. These structural choices have become popular in contemporary psychological thrillers, as they effectively create tension and maintain reader interest throughout the novel.

Here are examples of different structures used in seven novels:

Three-Act Structure: "Pride and Prejudice" by Jane Austen

Act 1: Elizabeth and Darcy meet and form initial impressions
Act 2: Complications arise as Elizabeth and Darcy's relationship faces obstacles and misunderstandings

Act 3: The resolution, where Elizabeth and Darcy overcome their pride and prejudice, leading to their happy ending.

Non-Linear Narrative: "The Time Traveler's Wife" by Audrey Niffenegger

The story jumps back and forth in time, following the relationship between Clare and Henry, who has a genetic disorder causing him to time travel unexpectedly.

Epistolary Novel: "The Perks of Being a Wallflower" by Stephen Chbosky

The story is told through a series of letters written by the protagonist, Charlie, to an anonymous recipient.

Frame Narrative: "The Name of the Wind" by Patrick Rothfuss

The main story is told as a flashback, with the protagonist, Kvothe, recounting his life to a chronicler

Parallel Narratives: "Cloud Atlas" by David Mitchell

The novel features six interconnected stories set in different time periods, with characters and themes that echo across the narratives.

Reverse Chronology: "All the Missing Girls" by Megan Miranda

The story is told backward, starting from Day 15 and moving back to Day 1, as the protagonist tries to unravel the mystery of her missing friend.

In Medias Res: "The Odyssey" by Homer

The epic poem begins in the middle of Odysseus's journey, with earlier events revealed through flashbacks and storytelling.

"When Harry Met Sally," "Star Wars," "Casablanca," and "The Wizard of Oz":

"When Harry Met Sally" - **Non-Linear Narrative with Elements of Three-Act Structure**

The story is told through a series of non-linear vignettes spanning over a decade, showing the evolution of Harry and Sally's relationship

Act 1: Harry and Sally meet, establishing their characters and initial impressions

Act 2: They become friends and navigate various relationships and life changes

Act 3: Harry and Sally realise their feelings for each other, leading to the resolution of their love story

. . .

"STAR WARS: A NEW HOPE" - **Hero's Journey and Three-Act Structure**

The story follows Luke Skywalker's hero's journey, a common structure in mythic tales.

Act 1: Luke's ordinary world is disrupted when he meets Obi-Wan Kenobi and learns about his destiny.

Act 2: Luke joins the rebellion, faces challenges, and rescues Princess Leia.

Act 3: The climactic battle against the Death Star and the resolution of Luke's initial journey.

"Casablanca" - **Three-Act Structure with Flashbacks**

The story is primarily linear but uses flashbacks to reveal the history between Rick and Ilsa

Act 1: Rick's life in Casablanca is introduced, and Ilsa arrives with her husband, Victor

Act 2: The complications of Rick and Ilsa's past relationship emerge, as does the political intrigue surrounding Victor

Act 3: Rick makes a decision that resolves his personal dilemma and contributes to the greater good

"The Wizard of Oz" - **Hero's Journey and Episodic Structure**

Dorothy's journey follows the hero's journey structure, with her adventure in Oz serving as the main narrative

The story is episodic, with Dorothy encountering various characters and challenges in each new location (Munchkinland, the Scarecrow's field, the Tin Man's forest, etc.)

The resolution comes when Dorothy defeats the Wicked Witch, meets the Wizard, and ultimately realises she had the power to return home all along

These examples show how classic films can combine different structural elements to create engaging and memorable narratives that have stood the test of time.

Some Examples of Tropes:

These are just a few:

Love Triangle: A romantic situation involving three characters, often leading to conflict and tough choices.

The Chosen One: A character with a special destiny or unique abilities who is destined to save the world or fulfill a significant role.

Coming-of-Age: Focuses on the growth, maturation, and self-discovery of a young protagonist.

Fish out of Water: A character who finds themselves in an unfamiliar or uncomfortable environment, often leading to humorous or insightful situations.

Redemption Arc: A character with a troubled past or flaws seeks forgiveness or atonement, leading to personal growth and change.

Forbidden Love: A romance between characters who face significant barriers or obstacles that prevent them from being together.

The Mentor: A wise and experienced character who guides and imparts wisdom to the protagonist.

Damsel in Distress: A female character who needs to be rescued or protected by a male character.

Unreliable Narrator: A narrator whose credibility or perception of events is compromised, often leading to a twist or revelation.

Rags to Riches: A character from humble origins who achieves wealth, success, or social status.

The MacGuffin: An object, goal, or device that drives the plot and motivates characters but may not hold much significance itself.

The Femme Fatale: A seductive and mysterious female character who often leads male characters into danger or ruin.

Battle of Good vs. Evil: The conflict between forces representing morality, righteousness, and evil, often with a clear distinction between heroes and villains.

The Unrequited Love: A one-sided love or affection that is not reciprocated by the other person.

The Twist Ending: A surprising and unexpected turn of events that changes the readerâ€™s understanding of the story.

The Double Cross: Betrayal or deception by a character who was trusted, leading to a significant plot twist.

The Tragic Hero: A protagonist with admirable qualities who ultimately meets a tragic fate due to their own flaws or circumstances.

Deus Ex Machina: A sudden and improbable resolution to a conflict or problem, often introduced by an unexpected external force.

The Underdog: A character or group that is underestimated or disadvantaged but overcomes obstacles to achieve success.

A writer can also have fun subverting tropes.

Gothic Romance Tropes

Here are some common tropes often found in Gothic romance:

The Haunted Castle or Mansion: The story takes place in a dark, eerie, and often dilapidated castle or mansion with a mysterious or tragic history.

The Brooding Hero: The male protagonist is typically a brooding, enigmatic figure with a dark past or secret. He is often wealthy, powerful, and initially cold or aloof toward the heroine.

The Curious Heroine: The female protagonist is an inquisitive, strong-willed woman who finds herself drawn to the mysterious hero and the secrets surrounding him. She often begins to uncover the dark history of the setting and the hero's past.

Supernatural Elements: Gothic romances often include supernatural or unexplained phenomena, such as ghosts, curses, prophecies, or eerie coincidences that contribute to the story's atmosphere of mystery and dread.

Dark Secrets and Mysteries: The plot revolves around unraveling secrets, often related to the hero's past, the history of the setting, or a mystery surrounding the heroine's own background.

Forbidden or Dangerous Love: The relationship between the hero and heroine is often forbidden, taboo, or socially unacceptable, adding to the story's tension and the characters' internal conflicts.

The Gothic Atmosphere: The setting and tone of the story are characterized by a sense of darkness, gloom, and impending danger. This atmosphere is often enhanced by descriptions of the architecture, weather, and the characters' emotions.

The Villain: Gothic romances often feature a villain who poses a threat to the heroine or stands in the way of the hero and heroine's relationship. This villain may be a rival suitor, a family member, or a supernatural entity.

The Damsel in Distress: Though Gothic heroines are often strong-willed, they may find themselves in situations where they need to be rescued by the hero, either from physical danger or from the threats posed by the story's villain.

The Romance Amidst Chaos: Despite the dark and often threatening circumstances surrounding them, the hero and heroine's love story remains at the heart of the Gothic romance, with their relationship developing and strengthening as they face the challenges and mysteries together.

These tropes, when used effectively, can create an atmospheric, suspenseful, and emotionally charged story that keeps readers engrossed in the world of Gothic romance.

Murder Mystery Tropes

Here are ten popular tropes and themes often found in the murder mystery genre:

The Locked Room Mystery: A crime, usually a murder, is committed under seemingly impossible circumstances, such as in a locked room with no apparent means of entry or escape.

The Unlikely Detective: The protagonist is often an amateur sleuth, such as a curious old lady (Miss Marple), a quirky private detective (Sherlock Holmes), or a nosy neighbor.

The Red Herring: False clues or suspects are presented to mislead the detective and the reader, adding complexity to the mystery and keeping readers guessing until the end.

The Gathering of Suspects: Near the end of the story, the detective often gathers all the suspects in one room to reveal the culprit and explain their deductive process.

The Least Likely Suspect: The guilty party is often the person who seems the least suspicious, subverting the reader's expectations.

The Motive Reveal: The detective uncovers the motive behind the crime, which is often rooted in human emotions such as greed, jealousy, revenge, or love.

The Dark Secret: The victim or the suspects often have a hidden past or secret that comes to light during the investigation, providing context for the crime.

The Isolated Setting: Many murder mysteries take place in isolated or confined settings, such as a remote mansion, a train, or a cruise ship, limiting the pool of suspects and creating a sense of tension.

The Sidekick: The detective often has a less-skilled assistant or friend who serves as a sounding board for their ideas and helps to expose the detective's thought process to the reader.

The Final Twist: Even after the culprit is revealed, there is often one last surprise or twist that catches the reader off guard and ties up any remaining loose ends.

THESE TROPES and themes have been popular in the murder mystery genre for decades, from classic authors like Agatha Christie and Arthur Conan Doyle to contemporary writers like Louise Penny and Tana French. While some may be considered clichés, skilled authors can still use these elements in fresh and innovative ways to create engaging and suspenseful stories.

Romance - Popular Tropes

If you've decided on a genre you wish to write, you can then drill down and find a range of tropes specific to this genre.

Bear in mind, also, that some tropes rise and fall in popularity while there are some tropes that ardent readers of the genre will *only* read.

Yes, I have had many readers tell me that their favourite trope is the 'single dad' or 'secret baby' or 'enemies to lovers' trope and that they don't care even who the writer is, they will consume everything they can that is written with that trope.

Here are some of the Romance Genre's most popular tropes:

Enemies to Lovers: Two characters start off with a strong dislike or rivalry but eventually develop romantic feelings for each other.

Friends to Lovers: A romantic relationship develops between two characters who have been friends for a long time.

Second Chance Romance: Former lovers reunite after a

period of separation or past difficulties, giving their relationship another chance.

Fake Relationship: Two characters pretend to be in a romantic relationship for various reasons, but their feelings eventually become genuine.

Forbidden Love: The romance faces significant obstacles and challenges due to societal, cultural, or familial constraints.

Opposites Attract: Two characters who have contrasting personalities, interests, or backgrounds find themselves drawn to each other romantically.

Love Triangle: A central character must choose between two potential love interests, creating tension and conflict.

Amnesia Romance: One character loses their memory and must rediscover their past, including their romantic relationship with another character.

Small Town Romance: The story takes place in a close-knit community where characters find love and romance amidst the charm and quirks of small-town life.

Age Gap Romance: The romantic relationship involves a significant difference in age between the characters, exploring the dynamics and challenges of such a connection.

Marriage of Convenience: Two characters enter into a marriage or partnership for practical or strategic reasons, only to develop genuine feelings for each other.

Slow Burn: The romantic relationship develops gradually over time, with intense chemistry and tension building up before the characters finally get together.

Love at First Sight: The characters experience an instant and powerful attraction to each other, often leading to a whirlwind romance.

Single Dad: This trope features a widowed or divorced father who must navigate the challenges of raising children alone while often finding new love.

Billionaire Romance: In this trope, one of the main characters is a wealthy and powerful individual, often leading to a Cinderella-like love story.

Science Fiction - Popular Tropes

Here are ten popular tropes within the science fiction genre:

Alien Invasion: Extraterrestrial beings arrive on Earth with the intention of conquering or destroying human civilization.

Time Travel: Characters journey through time, often to change the course of history or prevent a future catastrophe.

Dystopian Future: Society is portrayed as oppressive, totalitarian, or post-apocalyptic, often as a cautionary tale.

Space Opera: Grand adventures set in space, featuring interstellar travel, galactic empires, and epic conflicts.

Artificial Intelligence: Sentient machines or robots that either coexist with or turn against their human creators.

Cyberpunk: A subgenre that explores the intersection of

advanced technology, cybernetics, and societal decay in a near-future setting.

First Contact: The initial encounter between humans and extraterrestrial life, often exploring the cultural, scientific, and philosophical implications.

Alternate History: Stories that imagine how the world might have developed if key historical events had unfolded differently.

Superhuman Abilities: Characters with extraordinary powers, often the result of genetic mutations, technological enhancements, or psychic abilities.

Terraforming: The process of transforming an inhospitable planet into one suitable for human life, often exploring the challenges and consequences of planetary engineering.

Cozy Mystery Tropes

Cozy mysteries have enjoyed a surge in popularity over the past few years, attracting a growing audience of readers seeking a light-hearted, engaging escape from the stresses of everyday life.

The genre's appeal lies in its combination of charming settings, quirky characters, and intriguing puzzles that challenge readers to solve alongside the amateur sleuth protagonist.

Cozy mysteries offer a sense of comfort and familiarity, often set in tight-knit communities where the characters feel like old friends and the resolution of the mystery brings a satisfying sense of justice and order.

The rise of themed cozies, focusing on hobbies like baking, crafting, or gardening, has also contributed to the genre's growth, as readers enjoy exploring new passions or immersing themselves in worlds that align with their interests.

Additionally, the increasing demand for feel-good, uplifting content in the face of global uncertainties has made cozy mysteries an even more attractive choice for readers seeking a temporary retreat into a world where the good guys always win, and the bad guys get their just desserts.

Here are ten popular tropes often found in cozy mysteries:

Amateur Sleuth: The main character is usually not a professional detective but rather an ordinary person with a keen eye for detail and a knack for solving puzzles.

Small Town Setting: Cozy mysteries often take place in charming, picturesque small towns where everyone knows each other, and the crime is an unusual occurrence.

Quirky Characters: The supporting cast is filled with eccentric, colourful characters, each with their own unique personality traits and potential motives.

Themed Mysteries: Many cozy mysteries revolve around a specific theme or hobby, such as cooking, crafting, or bookstores, which the amateur sleuth is passionate about or involved in.

The Nosy Neighbour: There's often a curious, gossipy character who knows everything about everyone in town and provides the sleuth with valuable information or clues.

The Reluctant Police: Local law enforcement is often portrayed as either incompetent or reluctant to investigate the crime, leaving the amateur sleuth to take charge.

Red Herrings: Cozy mysteries are filled with false clues and misleading suspects to keep readers guessing until the end.

The Cozy Gathering: A significant event, such as a festival, reunion, or holiday celebration, brings characters together and sets the stage for the crime.

The Personal Connection: The amateur sleuth often has a personal stake in solving the crime, either because the victim was a friend or family member or because they've been implicated in the crime themselves.

The Dramatic Reveal: In the climax of the story, the sleuth gathers all the suspects together and reveals how they solved the mystery, often in a dramatic or unexpected fashion.

Historical Romance Tropes

Here are 10 popular tropes often found in historical romance novels.

I've written more than thirty historical romances and I have used every single one of these tropes in my various novels.

It has been so much fun!

THE MARRIAGE OF CONVENIENCE: The hero and heroine enter into a marriage for practical reasons, such as financial security or social status, but eventually fall in love with each other.

THE RAKE AND THE INNOCENT: A notorious rake (a man with a reputation for womanizing) falls for an innocent, virtuous woman who reforms him through the power of love.

THE WALLFLOWER: A shy, overlooked, or socially awkward heroine catches the eye of a dashing hero who appreciates her unique qualities.

• • •

THE FORBIDDEN LOVE: The hero and heroine are from different social classes, cultures, or have other barriers that make their love forbidden or taboo.

THE MASKED BALL: A chance encounter at a masked ball leads to a mysterious, romantic connection between the hero and heroine.

THE GOVERNESS AND THE LORD: A governess or other working-class heroine falls in love with her employer, a wealthy lord or aristocrat.

THE ARRANGED MARRIAGE: The hero and heroine are forced into an arranged marriage but eventually develop feelings for each other.

THE ENEMIES TO LOVERS: The hero and heroine start off as rivals or enemies but gradually develop romantic feelings as they spend more time together.

THE RESCUE ROMANCE: The hero saves the heroine from danger, such as a kidnapping or an unwanted suitor, and they fall in love in the process.

THE SECOND CHANCE ROMANCE: The hero and heroine have a history together, such as a past love affair or a failed engagement, and are reunited years later for a second chance at love.

THESE TROPES HAVE BEEN popular in historical romance for decades and continue to be used in contemporary novels set in various historical periods, from the medieval era to the Regency period and beyond. Authors often combine or subvert these tropes to create unique and engaging love stories that transport readers to another time and place.

Romantic Comedy - Popular Tropes

Here are ten popular tropes often found in romantic comedies:

Enemies to Lovers: The main characters start off disliking or even hating each other, but gradually develop romantic feelings as they spend more time together.

Fake Relationship: The protagonists pretend to be a couple for various reasons, such as to impress others or avoid unwanted attention, but eventually fall in love for real.

Friends to Lovers: Close friends realize they have developed romantic feelings for each other and must navigate the transition from friendship to love.

Love Triangle: The main character is torn between two potential love interests, often with contrasting personalities or representing different life paths.

Opposites Attract: Two characters with completely different personalities, lifestyles, or backgrounds find themselves drawn to each other despite their differences.

Misunderstandings and Miscommunication: The plot is driven by the characters' inability to express their true feelings or intentions, leading to humorous or dramatic misunderstandings.

The Bet or The Dare: One character makes a bet or accepts a challenge to win over or seduce the other, but genuine feelings develop along the way.

The Meet-Cute: The main characters' first encounter is charming, quirky, or memorable, setting the stage for their budding romance.

The Grand Gesture: One character goes to great lengths to prove their love or win back the other, often in a public or elaborate display of affection.

Second Chance Romance: Former lovers who parted ways due to misunderstandings, external circumstances, or personal growth find themselves reunited and given another chance at love.

"When Harry Met Sally" Genre & Trope Analysis

"When Harry Met Sally" is a romantic comedy film that follows the evolving relationship between the two main characters over the course of several years. The movie employs several tropes and conventions of the romantic comedy genre.

One of the central tropes is the "friends-to-lovers" storyline, where two characters who initially have a platonic relationship gradually develop romantic feelings for each other. Harry and Sally's journey from acquaintances to close friends and eventually to romantic partners is a quintessential example of this trope. The film also explores the question of whether men and women can be friends without romantic complications, a common theme in romantic comedies.

Another trope present in the movie is the use of contrasting personalities. Harry is pessimistic and cynical about love, while Sally is optimistic and believes in true love. Their differing viewpoints lead to witty banter and humorous disagreements, a staple of the romantic comedy genre. The film also features the "grand gesture" trope, where Harry publicly declares his love for Sally on New Year's Eve, a pivotal moment in their relationship. "When Harry Met

Sally" is considered a classic romantic comedy that both adheres to and subverts the genre's conventions, paving the way for future films in the genre.

Contemporary Romance Tropes

Here are ten tropes that are popular in contemporary romance:

Friends to Lovers: The hero and heroine start off as close friends but gradually develop romantic feelings for each other.

Second Chance Romance: The protagonists have a history together, such as a past relationship or missed opportunity, and are given a second chance at love.

Fake Relationship: The hero and heroine pretend to be in a relationship for various reasons (e.g., to impress family or to secure a business deal) but end up falling for each other for real.

Enemies to Lovers: The protagonists start off disliking or even hating each other but eventually develop romantic feelings as they spend more time together.

Billionaire Romance: A wealthy, powerful hero (often a billionaire businessman) falls in love with an ordinary or working-class heroine.

Small Town Romance: The story is set in a charming, close-knit small town where everyone knows each other, and the hero and heroine find love amidst the community.

Workplace Romance: The hero and heroine work together

and fall in love despite professional challenges or office policies against relationships.

Single Parent Romance: One or both of the protagonists are single parents, and their love story unfolds while they navigate the challenges of parenthood.

Opposites Attract: The hero and heroine have very different personalities, backgrounds, or lifestyles but find themselves drawn to each other despite their differences.

Forced Proximity: The hero and heroine are forced to spend time together due to circumstances beyond their control (e.g., being stranded together or sharing living quarters), leading to the development of romantic feelings.

These tropes are often combined or subverted in contemporary romance novels to create unique and engaging love stories that resonate with readers' experiences and fantasies.

Thriller Tropes

Here are 10 popular tropes often found in the thriller genre:

The Ticking Clock: The protagonist is racing against time to stop a catastrophic event, solve a mystery, or save someone's life.

The Conspiracy: The main character uncovers a wide-reaching conspiracy involving powerful people or organisations, often putting their own life in danger.

The Unreliable Narrator: The story is told from the perspective of a narrator who may be lying, hiding information, or whose perceptions are distorted, keeping the reader guessing.

The MacGuffin: The plot revolves around a sought-after object, piece of information, or person that drives the characters' actions and motivations.

The Femme Fatale: A seductive and manipulative female character who uses her charms to deceive or ensnare the protagonist.

The Innocent Bystander: An ordinary person is caught up in extraordinary circumstances and must rise to the challenge to survive or uncover the truth.

The Twist Ending: A sudden, unexpected revelation at the end of the story that changes the reader's understanding of the events that transpired.

The Cat-and-Mouse Game: The protagonist and antagonist engage in a tense, back-and-forth pursuit, trying to outwit and outmaneuver each other.

The Moral Dilemma: The main character must make a difficult choice between two equally unfavorable options, often with far-reaching consequences.

The Psychological Manipulation: The antagonist uses psychological tricks, gaslighting, or mind games to control or mislead the protagonist and other characters.

These tropes have been used effectively in various subgenres of thrillers, including psychological thrillers, spy thrillers, legal thrillers, and action thrillers. Authors often combine or subvert these tropes to create unique and gripping narratives that keep readers on the edge of their seats.

"Gone with the Wind" - Genre, Trope and Theme analysis

"Gone with the Wind" by Margaret Mitchell is a historical romance novel set in the American South during the Civil War and Reconstruction era. The novel incorporates several genres and tropes, most notably historical fiction, epic romance, and the Southern Gothic tradition.

As a work of historical fiction, "Gone with the Wind" offers a detailed portrayal of the antebellum South, the Civil War, and the Reconstruction period. Mitchell's vivid descriptions of plantation life, the horrors of war, and the societal changes that followed the Confederacy's defeat provide a backdrop for the central story. The novel also employs the trope of the Southern belle in the character of Scarlett O'Hara, who embodies both the charm and the flaws of the archetype.

The epic romance between Scarlett and Rhett Butler is another central theme in the novel. Their tumultuous relationship, marked by passion, pride, and misunderstandings, spans the course of the story and reflects the changing world around them. The novel also incorporates elements of the Southern Gothic tradition, which often explores the darker aspects of Southern culture and history. This is evident in the portrayal of the South's decline, the moral ambiguity of characters like Scarlett and Rhett, and the haunting presence of Tara, the O'Hara family plantation.

"Gone Girl" - 'Unreliable Narrator' & other Tropes

"Gone Girl" by Gillian Flynn is a psychological thriller novel that subverts and challenges many tropes commonly associated with the genre. The story follows the disappearance of Amy Dunne and the subsequent investigation that implicates her husband, Nick, as the prime suspect.

One of the central tropes Flynn employs is the unreliable narrator. The novel alternates between Nick's perspective and Amy's diary entries, both of which offer conflicting and misleading accounts of their relationship and the events surrounding Amy's disappearance. This narrative structure keeps readers guessing and questioning the truth, a key element of the psychological thriller genre.

Another trope Flynn subverts is the idea of the perfect victim. Initially, Amy is portrayed as a loving, devoted wife who has gone missing, garnering sympathy from the public. However, as the story unfolds, it becomes clear that Amy is far more complex and manipulative than she appears, challenging the notion of the innocent victim and blurring the lines between protagonist and antagonist.

The novel also plays with the concept of the media's influence on public perception, a common theme in contemporary thrillers.

As the investigation into Amy's disappearance unfolds, the media's portrayal of Nick and the couple's marriage shifts dramatically, showcasing how public opinion can be swayed by carefully crafted narratives.

Overall, "Gone Girl" takes familiar tropes of the psychological thriller genre and twists them in unexpected ways, creating a gripping and unsettling narrative that keeps readers engaged until the very end.

Gothic Romance Tropes

Here are some common tropes often found in Gothic romance:

The Haunted Castle or Mansion: The story takes place in a dark, eerie, and often dilapidated castle or mansion with a mysterious or tragic history.

The Brooding Hero: The male protagonist is typically a brooding, enigmatic figure with a dark past or secret. He is often wealthy, powerful, and initially cold or aloof toward the heroine.

The Curious Heroine: The female protagonist is an inquisitive, strong-willed woman who finds herself drawn to the mysterious hero and the secrets surrounding him. She often begins to uncover the dark history of the setting and the hero's past.

Supernatural Elements: Gothic romances often include supernatural or unexplained phenomena, such as ghosts, curses, prophecies, or eerie coincidences that contribute to the story's atmosphere of mystery and dread.

Dark Secrets and Mysteries: The plot revolves around unraveling secrets, often related to the hero's past, the history of the setting, or a mystery surrounding the heroine's own background.

Forbidden or Dangerous Love: The relationship between the hero and heroine is often forbidden, taboo, or socially unacceptable, adding to the story's tension and the characters' internal conflicts.

The Gothic Atmosphere: The setting and tone of the story are characterized by a sense of darkness, gloom, and impending danger. This atmosphere is often enhanced by descriptions of the architecture, weather, and the characters' emotions.

The Villain: Gothic romances often feature a villain who poses a threat to the heroine or stands in the way of the hero and heroine's relationship. This villain may be a rival suitor, a family member, or a supernatural entity.

The Damsel in Distress: Though Gothic heroines are often strong-willed, they may find themselves in situations where they need to be rescued by the hero, either from physical danger or from the threats posed by the story's villain.

The Romance Amidst Chaos: Despite the dark and often threatening circumstances surrounding them, the hero and heroine's love story remains at the heart of the Gothic romance, with their relationship developing and strengthening as they face the challenges and mysteries together.

These tropes, when used effectively, can create an atmospheric, suspenseful, and emotionally charged story that keeps readers engrossed in the world of Gothic romance.

The "Fairytale Retelling" Trope

The "fairytale retelling trope" has gained significant popularity in recent years, captivating audiences across various media, including literature, film, and television. This trope involves taking classic fairy tales and reimagining them with fresh perspectives, often updating them for modern audiences or exploring the stories from different angles.

One reason for the popularity of fairytale retellings is the familiarity and nostalgia associated with these classic stories. Many people grow up hearing fairy tales, and the opportunity to revisit these beloved narratives with a new twist can be both comforting and exciting. Retellings allow audiences to engage with familiar characters and plots while discovering new themes, interpretations, and insights.

Another factor contributing to the popularity of this trope is the potential for subversion and social commentary. Retellings often challenge the traditional values and gender roles present in classic fairy tales, offering a more progressive or inclusive perspective. For example, retellings might feature strong, independent female characters who reject the notion of waiting for a prince to save them, or they might explore the complexities of villains, providing them with more nuanced backstories and motivations. By reimagining these stories, authors and filmmakers can address contemporary issues and spark discussions about social norms, equality, and representation.

Moreover, the versatility of the fairytale retelling trope allows for endless creative possibilities. These stories can be adapted to fit various genres, such as fantasy, romance, horror, or science fiction, attracting a wide range of audiences. Retellings can also be set in different time periods or cultures, adding fresh elements to the original tales and exploring universal themes in new contexts.

As long as there are new perspectives to explore and creative twists to uncover, the appeal of this trope is likely to continue in the years to come.

Recent Tropes You May Never Have Heard Of!

In recent years, the romance genre has seen the emergence of a number of unconventional subgenres and tropes that have gained surprising popularity among readers.

Two notable examples are reverse harem and space romance.

Reverse harem, a subgenre that originated in Japanese manga and anime, features a female protagonist who is romantically pursued by three or more male love interests.

This trope subverts the traditional romance dynamic and explores themes of female empowerment, as the heroine navigates complex relationships and personal growth.

The popularity of reverse harem has been attributed to its escapist fantasy elements, diverse cast of characters, and the opportunity for readers to vicariously experience being desired and cherished by multiple partners.

Space romance, on the other hand, combines elements of science fiction and romance, setting love stories in the vast expanse of outer space or on distant planets. This subgenre allows authors to explore unique worldbuilding, futuristic technologies, and the challenges of love in a universe filled with adventure and uncertainty.

Space romance appeals to readers who crave both the excitement of sci-fi and the emotional satisfaction of a happily-ever-after ending. The popularity of space romance has grown alongside the increasing mainstream acceptance of geek culture and the success of science fiction franchises like Star Wars and Guardians of the Galaxy.

Other unusual tropes that have gained traction in the romance genre include:

Monster romance: Love stories involving mythical creatures, such as vampires, werewolves, or dragons.

Bully romance: A controversial trope where the love interest initially bullies or antagonizes the protagonist before they develop romantic feelings.

Mafia romance: Stories featuring love interests who are part of organized crime families.

Reverse age gap: Romance novels where the heroine is significantly older than the hero.

Polyamory: Love stories that explore consensual non-monogamous relationships involving multiple partners.

The emergence of these unconventional subgenres and tropes demonstrates the romance genre's adaptability and its readers' appetite for fresh, diverse perspectives on love and relationships. As the genre continues to evolve, it's likely that we'll see even more innovative and surprising tropes capture readers' imaginations in the years to come.

What is Required for a Scene to Exist?

Hopefully, the previous detailed discussion of tropes, and the many trope examples listed, may have sharpened your enthusiasm, and helped you refine your story idea.

So, let's return to the writing mechanics.

Regardless of tropes, your story is going to consist of scenes.

So, what is required for a Scene to Exist?

In fiction, a scene is a fundamental building block of storytelling. It is a self-contained unit of action, dialogue, and description that moves the plot forward, reveals character, and engages the reader. For a scene to exist and function effectively, it must have the following key elements:

1. **Purpose**: Every scene should have a clear purpose that contributes to the overall story. It should advance the plot, reveal character, provide important information, or create a turning point. If a scene doesn't serve a specific purpose, it may be unnecessary and should be cut or revised.

Example: In a romance novel, a scene where the protagonist meets her love interest for the first time serves the purpose of introducing a key character and setting up the central relationship of the story.

2. **Setting**: A scene must take place in a specific location and time. The setting should be described in enough detail to help the reader visualise the scene and understand its context. The setting can also contribute to the mood and tone of the scene.

Example: A suspense thriller might include a scene set in a dark, abandoned warehouse at night, creating a sense of danger and unease.

3. **Characters**: A scene must include at least one character, and often involves interaction between multiple characters. The characters' actions, dialogue, and internal thoughts should be consistent with their established personalities and motivations.

Example: In a mystery novel, a scene where the detective interrogates a suspect would involve a dynamic interaction between two characters with conflicting goals.

4. **Conflict or Tension:** A scene should have some level of conflict or tension to keep the reader engaged. This can be external, such as a physical confrontation or argument, or internal, such as a character grappling with a difficult decision or emotional turmoil.

Example: In a coming-of-age story, a scene where the teenage protagonist argues with their parents about a curfew creates conflict and highlights the character's struggle for independence.

5. **Dialogue and Action:** A scene should include a balance of dialogue and action to keep the pacing dynamic and the reader interested. Dialogue should sound natural and reveal character, while action should be descriptive and contribute to the scene's purpose.

Example: In a historical fiction novel, a scene at a royal banquet might include lively dialogue between nobles and servants, as well as descriptive action of the lavish food and entertainment.

6. **Sensory Details:** To fully immerse the reader in a scene, include sensory details that appeal to the five senses: sight, sound, smell, taste, and touch. These details help create a vivid and engaging experience for the reader.

Example: In a fantasy novel, a scene in a magical forest might describe the shimmering leaves, the melodic birdsong, the fragrant wildflowers, the taste of ripe berries, and the soft moss underfoot.

7. **A Clear Beginning and End:** A scene should have a definite starting point and a sense of resolution or transition at the end. The beginning should set up the scene's purpose and context, while the end should provide a sense of completion or lead into the next scene.

Example: A scene in a sci-fi adventure might begin with the crew of a spaceship receiving a distress call and end with them making the decision to investigate, leading into the next scene of their mission.

By incorporating these essential elements, writers can craft scenes that are engaging, purposeful, and integral to the overall narrative. Well-constructed scenes are the foundation of compelling fiction and keep readers invested in the characters and their journeys.

The 'High Concept' - & How to Nail it in One Paragraph

A 'high concept' is a concise, compelling description of your story's premise that immediately captures the attention of agents, publishers, and potential readers. It's the core idea that sets your book apart and makes it marketable.

When I was a cadet journalist on a country newspaper, one of my weekly jobs was to write the one-sentence description of the weekly television shows for the TV Guide.

At the time, I didn't realise what good experience it was for trying to distil my future 80,000-word stories into a compelling high concept.

To nail your high concept in one paragraph, follow these guidelines:

1. **Start with a hook:** Begin your paragraph with a strong, attention-grabbing statement that introduces the central conflict or unique aspect of your story. This could be a surprising situation, an intriguing question, or a bold claim.

Example: "In a world where dreams are the ultimate form of entertainment, one man's nightmares hold the key to saving humanity."

2. **Introduce your protagonist:** Briefly describe your main

character and the key characteristics or skills that make them interesting or well-suited to tackle the central conflict.

Example: "Jenna, a brilliant but socially awkward computer programmer, discovers she has the ability to manipulate reality through her code."

3. **Establish the stakes:** Clearly convey what's at risk for your protagonist and why the central conflict matters. This could be a personal stakes, such as the loss of a loved one or a threat to the protagonist's way of life, or broader stakes that impact society or the world.

Example: "When an ancient artifact unleashes a deadly plague, archaeologist Liam Carter must race against time to find a cure before it wipes out humanity."

4. **Hint at the obstacles:** Give a sense of the challenges or opposition your protagonist will face in their journey. This could be a powerful antagonist, a ticking clock, or a moral dilemma.

Example: "In a city where superpowers are outlawed, a young woman with telekinetic abilities must evade a ruthless government agency while unraveling a conspiracy that threatens to destroy everything she holds dear."

5. **Keep it concise:** Your high concept should be no more than one or two sentences, as it's meant to be a quick, impactful pitch. Avoid getting bogged down in details or subplots; focus on the core idea that makes your story unique and engaging.

By crafting a compelling high concept paragraph, you'll be better equipped to pitch your story and generate interest from industry professionals and readers alike. Remember, your high concept should be a distillation of your story's most exciting and marketable elements, so take the time to refine and perfect it.

Here's an example of a complete high concept paragraph:

"In a near-future world where virtual reality has replaced human interaction, a lonely teenager discovers a hidden online community that offers the promise of genuine connection. But as she becomes increasingly immersed in this digital utopia, she realises the community harbours a sinister secret that threatens to consume her reality."

This high concept paragraph introduces the hook (a hidden online community in a VR-dominated world), the protagonist (a lonely teenager), the stakes (the threat of losing touch with reality), and the obstacle (a sinister secret within the community), all in a concise and intriguing manner.

High Concept Examples of Popular Books & Films

Here are examples of high concept descriptions for the following:

1. **"When Harry Met Sally"**: Can men and women ever be just friends? Two cynical New Yorkers navigate a decade-long friendship that toes the line between platonic and romantic.

2. **"Gone with the Wind"**: A headstrong Southern belle must choose between her love for a roguish man and her loyalty to her family and homeland as the Civil War tears her world apart.

3. **"The Wizard of Oz"**: When a tornado transports a lonely Kansas farm girl to a magical world, she must navigate a strange land filled with whimsical characters and dangerous foes to find her way back home.

4. **"Gone Girl"**: When a woman disappears on her fifth wedding anniversary, her husband becomes the prime suspect as shocking secrets from their marriage come to light in a twisting psychological thriller.

5. **"Always Sunny in Philadelphia"**: Five self-centred friends navigate absurd misadventures and morally questionable schemes in their quest for personal gain at a failing Irish bar in Philadelphia.

6. **"Cinderella"**: A kind-hearted young woman, oppressed by her wicked stepmother and stepsisters, gets a chance at love and a better life when a fairy godmother grants her wish to attend a royal ball.

7. **"Carmen"**: A fiery and seductive gypsy woman entices a soldier to abandon his duties, setting off a chain of passion, jealousy, and tragic consequences in this iconic opera.

8 **"The Girl with the Dragon Tattoo"**: A disgraced journalist and a brilliant but troubled computer hacker unite to solve a decades-old disappearance while confronting corruption and violence in the dark underbelly of Swedish society.

9. **"Outlander"**: A World War II nurse is transported back in time to 18th-century Scotland, where she becomes entangled in a passionate romance with a Highland warrior amidst the tumultuous Jacobite Uprising.

10. **"Midsomer Murders"**: In a picturesque English county, a perceptive detective investigates a series of perplexing and sinister murders that disrupt the seemingly idyllic rural communities.

These high concept descriptions aim to capture the core premise, main characters, and central conflicts of each title in a concise and compelling manner, giving a clear sense of what makes each story unique and engaging.

How to Avoid Writing clichés

While genres come with certain expectations, the most compelling stories often find ways to innovate within those frameworks. Trust your unique voice and perspective, and don't be afraid to take risks and break the mould when it serves your story and characters.

Here are some ways you can avoid cliches in your genre fiction novel.

Subvert expectations: Take familiar tropes and put a fresh spin on them. Instead of relying on predictable plot points or character archetypes, find ways to surprise your readers by subverting their expectations. This can make your story feel original and engaging, even if it's working within a well-established genre.

Develop complex, multi-dimensional characters: Avoid flat, stereotypical characters by giving your protagonists and antagonists depth, unique motivations, and relatable flaws. Ensure that their actions and decisions are driven by their distinct personalities and backgrounds rather than genre conventions.

Explore new settings or time periods: Set your story in a unique or lesser-known location, or during a time period that hasn't been extensively explored in your genre. This can help your novel stand out and provide fresh opportunities for storytelling and character development.

Blend genres: Combine elements from different genres to create a unique and unpredictable story. For example, you could blend science fiction with romance, or horror with historical fiction. This cross-genre approach can help you avoid relying too heavily on the clichés of a single genre.

Focus on emotional truth and human experience: Rather than relying on plot devices or genre tropes, prioritise the emotional journey of your characters and the human experiences that drive their actions. By grounding your story in authentic, relatable emotions and conflicts, you can transcend genre clichés and create a powerful, resonant narrative.

The Value of Competitions - & How to By-pass the Slush Pile

As a new author, the idea of getting your work in front of an editor or agent can feel overwhelming and nearly impossible.

I'd had dozens of rejections before I joined the *Australian Romance Writers of Australia* (RWA) and began entering their competitions.

Thanks to the feedback of volunteer judges, I gradually improved my craft. RWA's competitions involve comprehensive feed-back, typically from two or three reader/writers judges for each entry.

Finally, I saw competitions as a way to by-pass the slush pile.

To increase my chances of catching the attention of agents and editors from my dream publishing houses, I strategically targeted writing competitions where editors and agents served as judges of the six finalists, hoping to land my work directly on their desks and potentially turn my publishing aspirations into reality.

This strategy ultimately paid off when the first three chapters of my historical romance, "The Courtesan's Daughter," were awarded first place by an editor from Harper Collins, who then requested the full manuscript. This experience taught me two valuable lessons.

Firstly, I had entered "The Courtesan's Daughter" in the

previous year's *Romance Writers of Australia* "First Three Chapters" Competition, where it ranked second to last and received disheartening feedback.

After allowing myself time to process the critiques, I carefully considered each piece of feedback from the three judges, rewrote the chapters, and re-entered the competition the following year.

To my surprise and delight, I won! This taught me the importance of perseverance, being open to constructive criticism, and continually working to improve my craft.

Secondly, winning the competition presented a new set of challenges. I had to finish the book within three months, and the required word count was 80,000 words. I dedicated myself to writing morning, noon, and night to meet the deadline. However, I found myself falling into old habits and making the same mistakes that had previously hindered my writing. I would establish the goal and conflict with a promising sub-plot that could be carried on in a series, but then the story would branch off into multiple sub-plots, losing focus and clarity.

It became clear that I needed to further refine my craft or work with a structural editor before submitting the full manuscript to Harper Collins. Unfortunately, I was young and inexperienced and I did not do this.

Ultimately, Harper Collins rejected the completed work. It was all part of the learning process. You don't know what you don't know.

And you don't know what you're doing wrong until someone points it out.

It was the editor who eventually commissioned my first published book who provided the valuable feedback I needed to get published.

"Your first half was great," he wrote. "You write great tension. But in the second half, you created three villains when you needed only one."

He advised me to rewrite and re-submit, which I did, and the book – "Lady Sarah's Redemption" – was published by Robert

Hale, UK, in 2009. It has since earned me a nice ongoing Public Lending Rights payment through its popularity in libraries throughout the Commonwealth.

Tips for Writing to a Particular Word Count

Writing to a specific word count can be a daunting task, whether you're crafting a short story, novella, or novel.

However, by employing a few key strategies and techniques, you can effectively organise your writing process and achieve your desired word count while creating a compelling and engaging narrative.

Outline your story

Before diving into the writing process, take the time to create a comprehensive outline of your story. This roadmap should include the main plot points, character arcs, and themes you want to explore. By having a clear structure in place, you can ensure that each scene and chapter serves a purpose and contributes to the overall narrative, helping you stay on track and avoid unnecessary tangents that can inflate your word count.

Determine your pacing

The pacing of your story will vary depending on your target word count and the conventions of your chosen genre. In a short story or novella, you'll need to be more concise, focusing on the most essential elements of your plot and character development. This may require you to be more selective in your descriptions and dialogue, choosing only the most impactful and relevant details. In a novel, you have more room to expand on your ideas, but it's still crucial to maintain a tight and focused narrative that keeps the reader engaged from beginning to end.

Set achievable writing goals

Break down your target word count into smaller, manageable goals. For example, if you're aiming to write a 50,000-word novel, you might set a daily goal of writing 1,000 words. By consistently meeting these smaller milestones, you'll make steady progress toward your overall word count while maintaining momentum and motivation.

Embrace the power of editing

Once you've completed your first draft, the editing process is essential for refining your story and achieving your desired word count. Review your work with a critical eye, identifying areas where you can trim excess language, eliminate redundancies, or cut scenes that don't serve the overall narrative. Be ruthless in your editing, asking yourself if each sentence and paragraph is truly necessary for the reader's understanding and enjoyment of your story.

Know your genre and market

Different genres and publishing markets have specific word count ranges that are considered standard or marketable. For example, a picture book typically falls between 500 and 700 words, while a young adult novel may range from 50,000 to 80,000 words. By researching and understanding these conventions, you can ensure that your story is well-suited for your intended audience and publishing goals.

Utilise writing tools and resources

There are numerous writing tools and resources available to help you stay organised and on track with your word count goals. Software programs like Scrivener or Ulysses offer features that allow you to set word count targets, track your progress, and easily rearrange scenes and chapters. Online writing communities, such as NaNoWriMo (National Novel Writing Month), provide support, encouragement, and accountability as you work toward your writing goals.

BY IMPLEMENTING these strategies and techniques, you can effectively write to a particular word count while crafting a compelling and engaging story. Remember, the key is to stay focused, organised, and committed to your writing goals, while also being open to feedback and willing to make revisions as needed. With dedication and persistence, you can create a polished and impactful narrative that resonates with readers and meets your publishing aspirations

How to Improve your Craft & Find Friends

Aspiring writers have numerous avenues to explore when seeking to improve their craft.

One of the most effective ways to grow as a writer is to join a writer's group. These groups provide a supportive community of fellow writers who offer feedback, encouragement, and accountability.

Writer's groups can be found locally, often through libraries, community centres, or bookstores, or online through platforms such as Facebook, Meetup, or Goodreads. Participating in a writer's group allows you to share your work, receive constructive criticism, and learn from the experiences of others. It also presents opportunities to network with other writers and potentially form lasting friendships and collaborations.

As mentioned, another valuable tool for beginner writers is entering writing competitions. Contests offer a chance to showcase your work, gain recognition, and potentially win prizes or publication opportunities.

They also provide deadlines and prompts that can motivate you to write regularly and explore new genres or styles. Many writing competitions are organised by local writing organisations, such as regional chapters of the Romance Writers of America, or the Romance Writers of Australia, Mystery Writers of America, or Science Fiction and Fantasy Writers of America.

These organisations often host workshops, conferences, and other events that provide further opportunities for learning and networking. For example, the Romance Writers of Australia's annual conference in August (generally rotating between Sydney, Melbourne, Adelaide, the Gold Coast, and Fremantle) offers an opportunity to pitch your manuscript to an agent or editor, in-person.

By researching and selecting competitions that align with your writing goals and interests, you can challenge yourself to improve your craft and gain exposure for your work.

*A list of resources, including places to submit to competitions, is included in a later chapter.

Pitching Your Book

Pitching any writing to a publisher requires careful attention to their requirements. If they accept unsolicited manuscripts, they'll generally state their requirements for submission. For example, it might be a short email query and nothing else.

It's also a good idea to see if a particular editor at that publishing house has a special interest in the kind of story you've written so that you can personalise your pitch.

Regardless, when pitching your genre fiction novel to an agent or publisher, it's essential to provide a concise and compelling package that showcases your work's potential. Traditionally, here are the key elements to include (though check the respective publisher or agent's guidelines):

1. **Query Letter:** Begin with a brief, engaging query letter that introduces your novel, its genre, and word count. Provide a short synopsis (1-2 paragraphs also known as the 'high concept') that captures the main plot, characters, and conflict. Mention any relevant writing credentials or awards, and explain why you believe your work is a good fit for the agent or publisher.

2. **Synopsis:** Include a more detailed synopsis (1-2 pages) that outlines the entire plot, including the resolution. This should give

the agent or publisher a clear understanding of your story's structure, pacing, and character development.

3. **Sample Chapters:** Provide the first three chapters or the first 50 pages of your manuscript, whichever is shorter. These sample chapters should be polished, engaging, and representative of your writing style and the overall quality of your work.

Different Options

When considering your publishing options, it's important to understand the different levels of traditional publishing and the alternative of indie publishing:

1. **"Big Five" Publishers:** These are the largest and most well-known publishing houses (Penguin Random House, HarperCollins, Macmillan, Simon & Schuster, and Hachette). They offer the widest distribution, extensive marketing support, and larger advances. However, they are highly selective and typically require authors to have an agent.

2. **Mid-sized and Small Presses:** These publishers may have a more niche focus and often accept submissions directly from authors without an agent. They offer more personalised attention but have smaller marketing budgets and distribution networks compared to the "Big Five."

3. **Indie Publishing:** This refers to self-publishing, where authors maintain complete control over the writing, editing, design, and marketing of their work. Platforms like Amazon's Kindle Direct Publishing, IngramSpark, and Draft2Digital enable authors to publish and distribute their books independently. While indie publishing offers greater creative control and higher royalties, authors are responsible for all aspects of the publishing process and must invest their own time and resources into marketing.

When deciding between traditional and indie publishing, consider your goals, target audience, and the resources you have available. Traditional publishing can provide validation, support, and wider reach, but it also involves giving up some control and earning lower royalties. Indie publishing allows for greater autonomy and faster time-to-market but requires authors to take on more responsibilities and financial risks.

Ultimately, the path you choose depends on your individual circumstances and priorities. Regardless of your publishing route, always strive to produce the highest quality work possible and be persistent in your efforts to connect with readers and build your author brand.

Vanity Publishing

Vanity publishing, also known as subsidy publishing, is a type of publishing where authors pay a company to publish their work. While vanity publishers may offer services similar to traditional publishers, such as editing, design, and distribution, there are several pitfalls authors should be aware of:

1. **High Costs:** Vanity publishers often charge substantial fees for their services, which can range from a few hundred to several thousand dollars. These costs can include editing, cover design, formatting, and a specific number of printed copies. Authors should carefully evaluate the value of these services and consider whether they can achieve similar results through indie publishing at a lower cost.

2. **Lack of Quality Control:** Vanity publishers typically do not have a rigorous selection process and will publish any book as long as the author pays the required fees. This lack of quality control can lead to the publication of subpar works, which can reflect poorly on the author and make it difficult to attract readers.

3. **Limited Distribution and Marketing:** Although vanity publishers may claim to provide distribution services, their reach is often limited compared to traditional publishers. They may not have

established relationships with bookstores, libraries, or other key outlets. Additionally, vanity publishers usually do not invest in significant marketing efforts, leaving authors responsible for promoting their own work.

4. **Lack of Industry Recognition:** Books published through vanity presses often lack the credibility and prestige associated with traditional publishing. Industry professionals, such as book reviewers, booksellers, and literary agents, may be less likely to take vanity-published books seriously, making it harder for authors to gain recognition and build a readership.

5. **Rights and Royalties:** Vanity publishing contracts may contain unfavorable terms regarding the rights to the author's work and royalty payments. Authors should carefully review any contract and seek legal advice if necessary to ensure they maintain control over their intellectual property and receive a fair share of any profits.

6. **Stigma and Reputation:** There is a stigma associated with vanity publishing, as it is sometimes seen as a last resort for authors who have been unable to secure a traditional publishing deal. This perception can make it more challenging for authors to be taken seriously and may hinder their ability to transition to traditional publishing in the future.

While vanity publishing may seem appealing to authors who are eager to see their work in print, it is essential to weigh the costs and potential drawbacks carefully. In many cases, indie publishing or pursuing traditional publishing through reputable channels may be more beneficial in the long run. Authors should research their options thoroughly, seek advice from experienced professionals, and make informed decisions based on their goals and resources.

Genre Competitions in Australia, US and UK

Australia

1. **Romance**: The Valerie Parv Award (Romance Writers of Australia)
2. **Horror**: The Australian Shadows Awards (Australian Horror Writers Association)
3. **Thriller**: The Ned Kelly Awards (Australian Crime Writers Association)
4. **Science Fiction/Fantasy:** The Aurealis Awards
5. **Young Adult:** The Inky Awards
6. **Historical Fiction:** The ARA Historical Novel Prize
7. **Crime Fiction:** The Davitt Awards and the Scarlet Stiletto Awards (Sisters in Crime Australia)

United Kingdom

1. **Romance**: The Romantic Novel Awards (Romantic Novelists' Association)
2. **Horror**: The British Fantasy Awards (British Fantasy Society)

3. **Thriller**: The Theakston Old Peculier Crime Novel of the Year Award

4. **Science Fiction/Fantasy:** The BSFA Awards (British Science Fiction Association)

5. **Young Adult:** The Branford Boase Award

6. **Historical Fiction:** The Walter Scott Prize for Historical Fiction

7. **Crime Fiction:** The Crime Writers' Association (CWA) Dagger Awards

United States

1. **Romance:** The RITA Awards (Romance Writers of America)

2. **Horror:** The Bram Stoker Awards (Horror Writers Association)

3. **Thriller:** The International Thriller Writers Awards

4. **Science Fiction/Fantasy:** The Hugo Awards (World Science Fiction Society)

5. **Young Adult:** The Michael L. Printz Award (American Library Association)

6. **Historical Fiction:** The Langum Prize for Historical Fiction

7. **Crime Fiction:** The Edgar Awards (Mystery Writers of America)

These genre-specific competitions offer writers the chance to showcase their work within their chosen genre and gain recognition from industry professionals and readers alike.

Keep in mind that some of these competitions may have specific entry requirements, such as membership in the organising association or specific publication dates. Always research the competition guidelines thoroughly before submitting your work.

Other Writing Competitions in Australia, US & UK

Australia

1. The Vogel Literary Award: A prestigious award for an unpublished manuscript by a writer under the age of 35.

2. The Australian/Vogel's Literary Award: An annual award for an unpublished manuscript by a writer under the age of 35.

3. The Richell Prize for Emerging Writers: An annual award for an unpublished manuscript by a writer who has not previously published a book-length work.

4. The Peter Porter Poetry Prize: An annual poetry prize open to all Australian poets.

5. The Calibre Essay Prize: An annual essay prize open to all Australian writers.

6. The Newcastle Short Story Award: An annual short story competition open to Australian writers.

7. The Scarlet Stiletto Awards: An annual crime and mystery short story competition for Australian women writers.

8. The Alan Marshall Short Story Award: An annual short story competition open to all Australian writers.

9. The Josephine Ulrick Literature Prize: An annual literature prize for short stories and poetry open to all Australian writers.

10. The Katharine Susannah Prichard (KSP) Writers' Centre Competitions: Various competitions throughout the year, including short story, poetry, and novella prizes.

United States

1. The National Book Awards: Annual awards celebrating the best books by U.S. authors in several categories, including fiction, nonfiction, poetry, and young people's literature.

2. The Pulitzer Prizes: Annual awards honoring excellence in journalism, literature, and musical composition, with categories including fiction, poetry, and drama.

3. The PEN America Literary Awards: A series of awards celebrating writers and writing across various genres, including fiction, nonfiction, poetry, and translation.

4. The National Poetry Series: An annual series of poetry competitions resulting in the publication of five books of poetry.

5. The Writer's Digest Annual Writing Competition: A series of competitions across various genres, including short stories, poetry, and essays.

6. The Pushcart Prize: An annual anthology featuring the best short stories, essays, and poetry published by small presses and literary magazines.

7. The O. Henry Prize: An annual collection of the best short stories published in U.S. and Canadian magazines.

8. The Flannery O'Connor Award for Short Fiction: An annual award for a collection of short stories by an American author.

9. The Iowa Short Fiction Award: An annual award for a collection of short stories by an emerging writer.

10. The Graywolf Press Nonfiction Prize: An annual award for a work of nonfiction by an emerging writer.

United Kingdom

1. The Bridport Prize: An annual competition for poetry, short stories, and flash fiction open to writers worldwide.

2. The BBC National Short Story Award: An annual award celebrating the best British short stories.

3. The Costa Book Awards: Annual awards recognizing outstanding books by writers based in the UK and Ireland, with categories including novel, first novel, biography, poetry, and children's book.

4. The Women's Prize for Fiction: An annual award celebrating excellence, originality, and accessibility in women's writing.

5. The Booker Prize: A prestigious annual award for the best novel written in English and published in the UK or Ireland.

6. The Commonwealth Short Story Prize: An annual award for the best piece of unpublished short fiction by a writer from a Commonwealth country.

7. The Seán Ó Faoláin International Short Story Competition: An annual short story competition open to writers worldwide.

8. The Moth Poetry Prize: An annual poetry prize open to all poets worldwide.

9. The Troubadour International Poetry Prize: An annual poetry competition judged by renowned poets.

10. The Bristol Short Story Prize: An annual short story competition open to all writers worldwide.

11. The Manchester Fiction Prize: A biennial prize celebrating excellence in creative writing.

12. The Deborah Rogers Foundation Writers Award: A biennial award for an unpublished writer who demonstrates outstanding literary talent.

13. The Mslexia Women's Novel Competition: An annual competition for unpublished novels by women writers.

14. The Mslexia Women's Short Story Competition: An annual competition for short stories by women writers.

15. The Times/Chicken House Children's Fiction Competition: An annual competition for unpublished children's novels.

. . .

THESE ARE JUST a few examples of the many writing competitions available in Australia and the United States. Aspiring writers can find numerous opportunities to showcase their work and gain recognition through these and other contests

Resources - Writers' Organisations

Australia has a vibrant writing community with numerous organisations supporting writers across various genres and regions. Here are some notable writer's organisations in Australia:

1. **Australian Society of Authors (ASA):** The ASA is a national organisation that represents the professional interests of authors across all genres, offering advocacy, advice, and support.

2. **Writers Victoria:** Based in Melbourne, Writers Victoria provides workshops, mentorship programs, and resources for writers at all stages of their careers.

3. **Writing NSW:** Formerly known as the NSW Writers' Centre, Writing NSW offers courses, workshops, and events for writers in New South Wales.

4. **Queensland Writers Centre (QWC):** QWC supports writers in Queensland through workshops, mentorships, and writing competitions.

5. **Writing WA:** This organisation serves writers in Western Australia, offering professional development opportunities, writing groups, and literary events.

6. **Writers SA:** Based in Adelaide, Writers SA provides support, resources, and opportunities for South Australian writers.

7. **TasWriters:** TasWriters is the peak organisation for writers in Tasmania, offering workshops, mentorships, and literary events.

8. **ACT Writers Centre:** Located in Canberra, the ACT Writers Centre supports writers in the Australian Capital Territory through workshops, events, and resources.

9. **Northern Territory Writers' Centre:** This organisation serves writers in the Northern Territory, offering workshops, writing groups, and literary events.

10. **Romance Writers of Australia (RWA):** The RWA supports writers of romantic fiction, offering conferences, workshops, and writing competitions.

11. **Australian Crime Writers Association (ACWA):** The ACWA represents writers of crime, mystery, and thriller fiction in Australia, offering networking opportunities and writing awards.

12. **Children's Book Council of Australia (CBCA):** The CBCA promotes Australian children's literature and supports writers and illustrators of children's books through events, awards, and advocacy.

13. **Australian Horror Writers Association (AHWA):** The AHWA supports writers of horror fiction in Australia, offering workshops, writing competitions, and networking opportunities.

14. **Australian Science Fiction and Fantasy Writers Association (ASFFWA):** The ASFFWA represents writers of science fiction and fantasy in Australia, offering writing awards, conventions, and professional development opportunities.

These organisations provide a wealth of resources, support, and community for Australian writers. Aspiring authors can benefit greatly from engaging with these organisations and participating in their events and programs.

Where to Find Writing Competitions

There are numerous online resources where writers can find information about various writing competitions. Here are some popular platforms and websites:

1. **Submittable** (https://manager.submittable.com/opportuni ties/discover): Many writing competitions and literary magazines use Submittable to manage their submission processes. You can search for open opportunities and filter by genre, entry fee, and deadline.

2. **Poets & Writers** (pw.org): This website maintains a comprehensive database of writing contests, grants, and awards. You can search by genre, entry fee, deadline, and prize amount.

3. **Winning Writers** (winningwriters.com): This online resource offers a directory of writing contests and markets, as well as articles and resources for writers.

4. **Reedsy** (blog.reedsy.com/writing-contests): Reedsy maintains a regularly updated list of writing competitions, categorized by genre and deadline.

5. **The Write Life** (thewritelife.com/writing-contests): This website publishes an annual list of writing contests, sorted by deadline and genre.

6. **Literary Magazines** (literarymagazines.com): This directory of literary magazines also includes a section for writing contests, searchable by genre and deadline.

7. **Writing.ie** (writing.ie/competitions): This Irish website maintains a list of writing competitions and opportunities, primarily focused on the UK and Ireland but also including some international contests.

8. **Australian Writers' Centre** (writerscentre.com.au/community/competitions): The Australian Writers' Centre maintains a list of writing competitions and awards open to Australian writers.

9. **New Zealand Society of Authors** (authors.org.nz/writing-opportunities): The NZSA website includes a section for writing opportunities, including competitions and awards open to New Zealand writers.

10. **Canadian Writers' Contest Calendar** (canadianwriterscontest.com): This website maintains a comprehensive list of writing contests and opportunities for Canadian writers.

Remember to always carefully review the guidelines, entry fees, and prizes for each competition before submitting your work. It's also essential to research the credibility and reputation of the organisation or publication hosting the contest to ensure it is a legitimate opportunity.

Tips for Polishing and Editing your work

Congratulations! You've finished your First draft.

Now the hard work begins!

Just kidding. Some writers complete a first draft that needs very little work because they edit and polish as they go.

However, this is the exception, especially if you are a less experienced writer.

Polishing and editing your work is an essential part of the writing process, and below are some suggestions to help get your manuscript ready to submit to an agent, publisher, or a competition.

Take a break: After finishing your first draft, set it aside for a while. Taking a break allows you to approach your work with fresh eyes and a clear mind.

Read aloud: Read your work aloud to yourself. This helps you identify awkward sentence structures, grammatical errors, and areas that need improvement. It also helps you gauge the flow and rhythm of your writing.

Focus on clarity and coherence: Ensure that your ideas are communicated clearly. Look for any confusing or convoluted sentences and rephrase them to enhance clarity. Ensure that your paragraphs flow logically from one to another, creating a cohesive and coherent piece.

ONCE I AM happy with the structure of my work as a whole, it's time to return to the specifics of the writing.

For this, I utilise my favourite free editing software—Pro-writing Aid. I paste my work into the programme so it can pick up typos, grammar mistakes, and suggest improvements to sentence structure which I can either accept or dismiss. Its suggestions can help with the following:

Cut unnecessary words and phrases: Eliminate wordiness and streamline your writing. Look for redundant words, clichés, and excessive adjectives or adverbs. Keep your sentences concise and to the point.

Check grammar, spelling, and punctuation: Pay close attention to grammar, spelling, and punctuation errors. Use spelling and grammar check tools, but don't rely solely on them. Proofread your work manually to catch any errors that might have been missed.

Vary sentence structure: Aim for sentence variety to maintain reader engagement. Alternate between short and long sentences, simple and complex structures, and vary your sentence beginnings to create a more dynamic writing style.

Develop strong transitions: Smooth transitions between paragraphs and ideas improve the overall flow of your writing. Use transitional words and phrases to connect thoughts and guide the reader through your work.

Seek feedback: Share your work with others, such as writing groups, friends, or trusted colleagues. Their fresh perspectives can provide valuable insights and constructive criticism to help you identify areas for improvement.

Consider the reader's perspective: Put yourself in the reader's shoes and evaluate your work from their standpoint. Is the information presented clearly? Are the ideas organised in a logical manner? Does the writing engage and hold their interest?

Revise and repeat: Editing is an iterative process. After making the necessary changes, revise your work again. Repeat this process until you are satisfied with the final result.

EDITING AND POLISHING take time and effort, but they are essential for transforming your initial draft into a polished, refined piece of writing.

Steps to Self-publish

- **Write your story**
- **Edit it**
- **Format it; and**
- **Publish it!**

These are the most basic steps required, but with so many options at each turn, it can be confusing to choose the best path for *you*.

Writing options

Some people write straight into Microsoft Word, others prefer Scrivener or other software to help organise their chapters.

PREVIOUS CHAPTERS in this book have detailed the writing process, but when you are happy with your manuscript and ready to publish, where do you go from there?

Publishing options

1. You can shop it around to find a traditional publisher.
2. You can pay a company to edit, format, publish (and, perhaps) market it for you.
3. You can do it all yourself (which requires a lot to learn).

Or - my preferred method -

4. You can outsource each different component—ie, editing, formatting, cover design, publishing—to different professionals while retaining overall control, much like you would project-manage a building project or kitchen renovation.

The positives and negatives of the traditional approach vs the do-it-yourself (Indie- or self-publishing) approach.

Pros of Signing with a Traditional Publisher:

Professional Editing and Production:

Traditional publishers typically have a team of professional editors, proofreaders, and designers who work on your book. They can help improve the quality of your writing, ensure proper formatting, and create an attractive cover design.

DISTRIBUTION AND MARKETING SUPPORT:

Traditional publishers have established distribution channels and relationships with bookstores, online retailers, and libraries. They can help get your book into physical and digital stores, increasing its visibility and accessibility to readers. Additionally, they often provide marketing support, including promotional campaigns, author events, and media exposure.

. . .

CREDIBILITY AND PRESTIGE:

Being published by a traditional publishing house can bring a level of credibility and prestige to your work. It is often seen as a validation of the quality and marketability of your writing, which can help build your reputation as an author.

ADVANCE AND ROYALTIES:

Traditional publishers may offer an advance payment to authors, which serves as an upfront payment before the book is published. Additionally, authors typically receive royalties based on book sales, allowing them to earn ongoing income from their work.

Drawbacks of signing with a traditional publisher

Limited Control:

When signing with a traditional publisher, authors may have to relinquish some control over their work. Publishers may have the final say on aspects such as editing, cover design, and title. The author's creative vision may not align completely with the publisher's decisions.

LONGER TIME TO PUBLICATION:

The traditional publishing process can be lengthy, involving several stages such as manuscript review, editing, production, and marketing. It may take years from the time of signing a contract to the actual release of the book.

LOWER ROYALTIES:

While traditional publishers offer advances and royalties, the percentage of royalties received by authors is often lower compared to self-publishing. This is because the publisher takes a significant share of the revenue to cover their costs, including editing, printing, marketing, and distribution.

LIMITED ACCEPTANCE AND COMPETITION:
Traditional publishers receive a vast number of submissions and can only accept a limited number of books each year. The competition to secure a publishing deal can be fierce, and many authors may face rejection or struggle to capture the attention of publishers.

Ultimately, the decision to sign with a traditional publisher depends on the individual author's goals, preferences, and circumstances. It is essential to carefully weigh these pros and cons, consider the specific terms and conditions of any publishing contract, and assess how well they align with your long-term writing and publishing objectives.

Pros and cons of paying a company to edit, format, publish and market for you.

If you don't go the traditionally published route but you don't have the energy or interest in self-publishing, then you might choose to pay someone, or a company, to do this for you.

BE CAREFUL. My biggest advice is to shop around and, if preferable, get word-of-mouth recommendations if you plan to hand your manuscript over to a publishing company to whom you will *pay* money, rather than who will pay *you*.

LOOK at the reviews about a company to whom you might be paying hundreds—of not thousands—of dollars to edit, format, and publish your work. The actual process of getting you work into print is not costly. (I will go through the costs of editing, cover design, and formatting in a later chapter.)

. . .

FIND out where their greatest costs lie? Are they promising advertising and promotional support? Remember that there can be no guarantees for the results of this support.

SO, check online reviews, or through a writers' organisation like the Alliance of Independent Authors (ALLi), a non-profit professional association for authors who self-publish. They have a global team who provide trusted, best-practice information and advice to the author community. Their Ask Alli forum answers any question anyone might have about self-publishing.

WHILE GETTING someone else to publish your book may be the ideal option for you, make sure they are a good fit.

Due Your Due Diligence and be Wary of Scams

As mentioned earlier in this book, novice writers who are eager to see their work in print can be vulnerable to scams when it comes to self-publishing or vanity publishing companies.

IT IS THEREFORE crucial to conduct thorough research, seek advice from reputable author communities or writing organisations, and carefully scrutinise any publishing agreements before committing time, money, and intellectual property.

SO, before making any binding agreements with a company to publish your work, check their reputation and ensure that they are reputable. The Alliance of Independent Authors is an organisation that can help you with this.

. . .

YOU MAY FIND the perfect partner to whom to hand over your work, and who will then produce a beautiful book. I've listed several in the Useful Resources and Software Chapter.

IF THEY ASK for extra money to market your book, be careful about what you agree to. You may decide you simply want them to produce the book which you can then upload to the retailers, retaining full control from that point onwards.

Choosing the Self-publishing Option and doing it yourself!

In the next chapter I'll go through the steps I use to go from finished manuscript to publishing my book in print and as an ebook.

Outsourcing the Components of your Project

This can seem more complicated than it is, largely because some of the steps need to be completed before others—if you are choosing one option over another.

FOR EXAMPLE, if I do my own formatting, I need to know the number of pages first, in order to inform the professional cover designer I've commissioned.

HOWEVER, if I choose to build my own cover using some of the tools within, for example Amazon's KDP or Ingram Spark, I don't need a wraparound cover, and therefore I can make changes to my manuscript without it having an impact on the cover.

OUTSOURCING the individual components is my preferred method as it offers me the most control for the cheapest price.

. . .

WHAT 'PARTS' can you outsource?

- Editing
- Cover
- Formatting your manuscript
- Uploading the correct files
- Printing

YOU'VE DONE THE HARD, time-consuming work of writing your novel. Now that you have finished it, you should feel very proud of what you've achieved!

GETTING a Word manuscript from your computer and into print or ebook can seem daunting.

LET'S go through the above, one by one.

Editing

You should always get someone to edit your work. You are too close to the words to get the big picture or to pick up the typos.

Below are options for editing:

1. Have a fellow writer/writer's group look at it. You want the story to be structurally sound. Or…
2. Find an editor to give it a structural edit. (An 80K manuscript might cost AUD400 to $1000 or $1500.)
3. You might not need a structural edit. A line edit/proofread might be sufficient.

Be aware that there are different types of editing.

A structural edit is 'big picture' editing.

A line/copy edit is like proofreading.

AFTER I'VE DECIDED that I'm happy with the structure and 'big picture' aspects of my work, I will put the manuscript through Pro-Writing Aid (as mentioned), which is software designed to pick up errors, and that suggests better writing, at the grammatical and typo stage.

YOU CAN GET a free version of this here:

https://prowritingaid.com/

Formatting

Formatting and uploading to the retailers costs little time and expense. As I mentioned earlier, you could outsource your manuscript in the same way you might project manage the building of a house or refurbishing of a kitchen. You will have control over which particular tiler, painter, stonemason, etc, you contract for the job.

Formatting is necessary to get the file formats you need for:

- Ebooks – (which requires an .epub)
- Print – (which requires two PDFs—one PDF for the interior, and one PDF for the cover).

How do you format your manuscript?

To format your file, two options I recommend are:

1. Vellum if you have a Mac. (https://vellum.pub/)
2. D2D (Draft2Digital) https://www.draft2digital.com/faq/

OR:

Outsource it to a formatter. This is a quick job and you should get it done for around $100.

I suggest you look for 'Book Formatter' on https://www.fiverr.com/ or at reedsy.com.

First, let me talk briefly about these two platforms which I have found so useful in my journey from writing the book to getting it into the world.

What is Fiverr?

Fiverr is a multinational online marketplace for freelance services. Its platform connects freelancers to people or businesses looking to hire. Listings on Fiverr are diverse and range from "get a well-designed business card" or "help with HTML, JavaScript, CSS, and jQuer" or "Find a Cover Designer" and "Find a Formatter".

What is Reedsy?

Reedsy is a platform that allows you to connect with designers, editors, translators, and marketers who have extensive experience in the publishing space.

Reedsy has freelancers for just about every step of the publishing process: there are editors and designers, translators, book marketers, website designers, book marketers, book reviewers, and even book publicists. Reedsy launched in November of 2014, making it an experienced and well-rounded marketplace.

Currently, there are over 150,000 authors who are using the platform. However, unlike most other freelancer platforms, every freelancer who receives an account is **vetted by the Reedsy team** to verify they perform their service at the highest level quality.

Now, back to the process of self-publishing ...

Once the interior of your book is formatted, you will need a book cover.

For an ebook, you only need the front cover. Sizing the cover for an ebook is not as important as getting the sizing exactly right for a print book.

WHEN YOU UPLOAD your files to get your book into print, print-on-demand publishers like Amazon, Ingram Spark, D2D, Lulu, etc, will give you the option to use your ebook cover as the front cover while you 'build' the back cover and spine, according to the trim size of your book, and the width of the spine.

YOU COULD MAKE your own cover using Canva, Bookbrush (designed for writers), or PicMonkey.

For an ebook, I only need the front cover

For a print book, the wraparound cover needs to incorporate the spine. (The spine width is dependent upon the number of pages.)

MY PREFERENCE IS to have a professional cover designer make me a wraparound cover.

FOR ONE OF my book series, I use Fiverr.com (and pay between AUD$45 and $100 per cover).

FOR ANOTHER SERIES, I use a US cover designer who understands the requirements of the historical romance genre in which I write. She was a word-of-mouth recommendation who makes many covers for genre writers. I pay her US$120 per cover.

What Information do you need to provide your cover designer when making a paperback book?

- Trim Size
- No. of pages
- Paper (cream or white)

Generally, fiction is printed on cream paper, while non-fiction is printed on white paper.

THE COVER DESIGNER needs to know whether white or cream paper is used because 100 pages of cream paper will result in a different spine width compared with 100 pages of white paper.

Use the Ingram Spark Template to produce the cover size to give to your designer

Alternatively, you can generate a template once you've uploaded your book's basic details. Then you'd simply email that to your cover designer. It's what I usually do.

ONCE YOU HAVE your cover back from the designer as a .PDF, you'll upload that, plus the interior .pdf file, to Ingram Spark or your chosen print-on-demand service or publisher.

Formatting the book yourself

This will need to be completed before you can get a wraparound cover sized correctly, because, as mentioned, the number of pages (and trim) will determine the exact dimensions.

ANOTHER OPTION for formatting yourself is to use Draft2Digital (D2D) – which is a multi-function platform.

THIS IS a great option if you have a Word document as you can have it formatted at no cost.

D2D will:

Format your eBook AND your print book.

IT WILL ALSO DISTRIBUTE your books to the retailers IF YOU CHOOSE.

. . .

HOWEVER, at no cost, you can feed in your Word file, and it will turn this into an .epub to upload it as an ebook, or .pdf which you will need if you plan to publish in print.

WITH THESE FILES, you are free to upload to the retail or printing or publishing platform of your choice.

SO, now that your book has been edited and formatted, what now?

PRINT

I use Ingram Spark for my print books because they have printers in Australia, as well as worldwide.

That means that if my book is being bought by someone in, for example, the UK, they will only pay local shipping because it will be printed at the closest local printer – on demand.

This makes it a cost-effective option for distribution, whether it is being acquired by a family member overseas, or a reader to whom I have sold my book.

OTHER PRINT-ON-DEMAND SERVICES INCLUDE LULU, Bookvault and Amazon. (These are the ones I use though there are many more.)

How to do it using Ingram Spark:

You can upload your two .pdf files to Ingram Spark (https://www.ingramspark.com/) and they will print and distribute it.

It's a good option for people who want small amounts of Print on Demand (POD) copies for family and friends.

You can set it to private or if you want to make it available commercially, and market to libraries and the public at large, just set it to public.

ON THE BELOW WEBSITES, you can prepare your book as both an ebook or a print book.

Here are the websites which will outline the specific instructions for doing either or both.

https://draft2digital.com/

https://www.ingramspark.com/

https://kdp.amazon.com/en_US/

Whichever sites you are going to use, you will have to set up a free account first.

THEN, if you want a print book, you will need two files:

- Interior
- Cover

Ingram Spark will require two already formatted .pdf files.

D2D IS SIMPLER as you can use a word .doc file for the interior.

NOTE!

. . .

FOR THE ABOVE PRINTERS, you don't *need* a single wraparound cover for your print book as they provide the tools to help you 'build' your own cover in three parts: Front, Spine, and Back.

HOWEVER, I choose to outsource a complete wraparound cover to my cover designer because I have no design talent while I know my cover designer will provide something much more professional than I could ever produce—and at a reasonable cost.

What do I need to Upload? At the risk of sounding repetitive, I'll say it again:

FOR INGRAM SPARK you will need:

- A formatted .pdf for the interior.
- A wraparound .pdf for the cover based on the number of pages of your interior.

TIP! The size of the cover of a print book will change with the addition of just a few pages as this alters the width of the spine for a wraparound cover.

Other useful Sites for Cover Design & Formatting

For formatting, I use Vellum for a .pdf interior.

https://vellum.pub/ (This is for Mac only)

https://www.fiverr.com/ A freelancer site that offers formatting, design, and many other related options https://reedsy.com/ - A great, curated site for finding editors, formatters, cover designers, etc

And for DIY design, Canva, PicMonkey, and BookBrush are great.

If I need to resize a cover my designer has created, I use Canva.

https://www.canva.com/en_au/
https://bookbrush.com/
https://www.picmonkey.com/

Draft2Digital - Formatting, Publishing & Distribution

Below are some of the frequently asked questions answered by D2D about their site and services.

D2D CHARGES nothing to use their tools. There are no costs associated with using them to format your book and generate the files you need to upload to any print-on-demand service.

HOWEVER, if you choose to distribute to the retailers through them, they will take a small percentage of each sale.

WHAT DO ***I need to make a Draft2Digital account?***

It is free to make an account with Draft2Digital. All you need to sign up is a name and an email address that you can check regularly. We will keep the name on your account separate from any Pen Names. There is no upfront cost to using Draft2Digital.

CAN ***you really turn my manuscript into a book?***

Yes! Draft2Digital offers ebook and print book publishing services. If you have a complete story or manuscript, D2D can convert your work into a beautiful electronic book (ebook) or print on demand (POD) paperback book.

As a self-published author, you are the publisher of your own books and you retain the intellectual property rights of your story. You control when and where Draft2Digital sends your book.

Even if you choose not to publish with us, we have you covered. Draft2Digital gives you the option of downloading the files we convert for you at no charge. You can create, convert, and download a pixel-perfect paperback PDF or ebook EPUB file without having to publish through D2D.

WHAT FILE TYPES ***can Draft2Digital take and make?***

The most popular file format that authors upload is a Word document. Upload your book in .doc or .docx file format, and we'll convert your manuscript into an ebook or print book. Anything Word can read, we can read. (RTF, OTF, TXT, too!)

If you already have an epub of your own, Draft2Digital accepts epub files for ebook, but not print. We won't make any changes to your epub formatting, but we'll gladly distribute it to all our digital stores for you.

WHAT FILE TYPES ***will Draft2Digital NOT accept?***

We do not accept printed manuscripts; Draft2Digital can only convert electronic file types. Because of the fixed nature of PDFs, Draft2Digital cannot convert PDFs to any other file type. PDFs are only accepted as print book interior or print book cover files.

WHERE'S YOUR STYLE GUIDE?

Our goal is to support *your* writing without you having to worry about formatting. As such, we accept a variety of document formatting styles and don't have any single style guide.

If you'd like some direction on creating a simple style that will convert beautifully every time, do these two things:

• Skip the title page, copyright page, dedication, also by, about the author, teaser, and any biographical pages. *You only have to upload your story.* Let us help you add the technical stuff later.

• Mark each new chapter or scene break with something distinctive, *and* ***be consistent***. Make new chapter titles centered and bold, or larger font, or use a heading style. The most important part is to do the same thing to each chapter title or scene break so that our system can learn your style.

That's enough for our tools to build a beautiful ebook out of your manuscript.

This all sounds great, but it's an automated system. What if there are problems with my book?

We've got you covered. Draft2Digital is famous for our friendly and knowledgeable customer support.

Ingram Spark

Ingram Spark talks about the following in their Frequently Asked Questions:

Editing

Some editors charge per word, some charge per page, and some charge per hour. The **Editorial Freelancer Association** provides rough guidelines to give you an idea of common editorial rates. (Note that these are US dollars.)

Developmental Editing (estimated pace 1-5 ms pgs/hr): US$45-55/hr

Substantive or Line Edit (estimated pace 1-6 ms pgs/hr): $40-60/hr

Basic Copyediting (estimated pace 5-10 ms pgs/hr): $30-40/hr

Proofreading (estimated pace 5-10 ms pgs/hr): $30-40/hr

When you're ready to hire an editor, do some research and

request a few different quotes. Meet with them, if possible, or have a conversation to get to know whether or not they'll be the best editor for your book's content. Have they edited in that genre before? Do you feel like they understand your writing style? Ask them for references or examples of their work to make the most informed decision.

ISBNS

ONE PUBLISHING EXPENSE you'll incur is purchasing an ISBN, if you choose not to use a free ISBN. This is necessary if you'd like to enable your book for distribution, and each format of each book you publish must have its own unique ISBN.

As part of Ingram Spark's catalog integrity initiative, they require that ISBNs are valid and properly acquired. Each country has a single approved and designated agency that issues ISBNs for publishers and self-publishers located in that country. Bowker is the official ISBN agency for the US; Nielsen is the official ISBN agency for the UK and Ireland; Thorpe-Bowker is the official ISBN agency for Australia. IngramSpark does business in many other countries, so they encourage each customer to ensure they are acquiring their ISBN from their country's approved ISBN Agency. You can find the approved ISBN agency for your country on the International ISBN Agency's Website.

For U.S. customers, ISBNs can be purchased from Bowker directly through your IngramSpark account for $85 each. You can also buy a block of ten ISBNs from Bowker for $295. IngramSpark believes it's in your best interest to be recognised as the owner of your work and a publisher in your own right, which is why they encourage publishers to purchase their own ISBNs.

If you want your book to be available in brick-and-mortar bookstores, you'll need a barcode in addition to your ISBN. Some ISBN services sell barcodes, as well as ISBNs, but you can use IngramSpark's book cover template generator to get a free barcode.

Printing & Shipping

The cost to print and ship a book with IngramSpark depends on the following:

- Trim Size
- Interior Colour and Paper
- Binding Type (paperback or hardback)
- Laminate Type (gloss, matte, or textured)
- Page Count
- Quantity
- Ship-To Address

INGRAMSPARK HAS a **Print and Ship Calculator** to help you understand how much it will cost.

Where to Publish your E-book? Kindle Unlimited or Wide?

When it comes to publishing ebooks, authors have two main options: enrolling in Amazon's exclusive Kindle Unlimited (KU) program or going "wide" by distributing their books through multiple retailers. Here's a breakdown of the differences, along with the pros and cons of each approach:

Kindle Unlimited (KU): KU is a subscription service offered by Amazon, where readers pay a monthly fee to access a vast library of ebooks. Authors who enrol their books in KU must grant Amazon exclusive distribution rights for a minimum of 90 days.

Pros:

1 Increased visibility: KU books are prominently featured on Amazon, which can lead to more downloads and exposure.

2 Borrowing incentives: Authors are paid per page read, which can potentially lead to higher earnings, especially for longer books or series.

3 Promotional tools: KU authors have access to Amazon's promotional tools, such as Kindle Countdown Deals and Free Book Promotions.

4 KU All-Star Bonuses: Top-performing KU authors can earn additional bonuses based on their page reads.

Cons:

1 Exclusivity: Authors cannot distribute their ebooks through other retailers while enrolled in KU.

2 Limited readership: Some readers prefer to purchase books outright or use other ebook platforms, limiting the potential audience for KU books.

3 Fluctuating income: Page read rates can change, and income may vary depending on reader engagement.

4 Shorter revenue tail: Once a book is removed from KU, it may experience a drop in visibility and income.

Going Wide

Going "wide" means distributing your ebook through multiple retailers, such as Amazon, Apple Books, Kobo, Barnes & Noble, and Google Play, among others.

Pros:

1 Diversified income: By not relying solely on Amazon, authors can potentially reach a wider audience and generate income from multiple sources.

2 Increased control: Authors have more control over pricing, promotions, and distribution of their books.

3 Long-term stability: Building a readership across multiple platforms can lead to a more stable and sustainable income over time.

4 Avoiding exclusivity: Authors can participate in promotions and bundle deals with other authors or retailers without being tied to Amazon.

Cons:

1 Reduced visibility on Amazon: Books that are not in KU may have lower visibility on Amazon's platform.

2 More complex management: Managing book distribution and promotions across multiple platforms can be time-consuming and require additional effort.

3 Lower royalties on some platforms: Some retailers offer lower royalty rates compared to Amazon, depending on the book's price.

4 Fewer promotional tools: Other platforms may have fewer built-in promotional tools compared to Amazon.

MY DECISION WAS to do both. I have several series that are in Kindle Unlimited, a decision I made because the eight books had good read-through. I put a buy link in the backmatter of each book, which often meant readers would automatically jump to read the next book, scoring me page reads.

For the rest of my books, which are wide, my person decision was to use the aggregator D2D to distribute my books to Apple, Barnes & Noble and multiple other retailers, libraries, etc.

However, I chose to upload directly to Kobo because of the promotions they offer, which are only available when you publish directly.

I also publish directly with GooglePlay.

At both Kobo and GooglePlay, I've uploaded my self-narrated audiobooks.

These decisions are all purely personal and based on the genre of books I write and where I've discovered my readership is. And that differs from series to series.

Ultimately, the choice between KU and going "wide" depends on an author's goals, target audience, and marketing strategy. Some authors find success with KU's exclusivity and promotional tools, while others prefer the flexibility and diversification of going "wide."

Many authors choose to experiment with both approaches to determine which works best for their specific situation.

There are many Facebook Groups that can help authors navigate their writing decisions and establish what is right for them or the genre in which they write.

Wide for the Win is one Facebook Group that has a wealth of practical information for authors who want to build their readership through platforms other than Amazon.

Useful Software for Writers

Here's a list of programs and software that can help writers with publishing or self-publishing their work, including Publisher Rocket for keyword research:

1 Scrivener: A popular writing software that helps authors organize their writing projects, research, and notes. It also provides formatting tools for exporting your work to various file types, including those suitable for self-publishing.

2 Vellum: A book formatting software for Mac that allows authors to create professionally designed ebooks and print books with ease. It offers a variety of templates and customisation options.

3 Draft2Digital: An ebook distribution platform that simplifies the process of publishing and distributing your book to multiple online retailers, including Amazon, Apple Books, Kobo, and Barnes & Noble.

4 Reedsy: A marketplace that connects authors with professional editors, designers, and marketers. They also offer a free book formatting tool called Reedsy Book Editor.

5 Calibre: A free, open-source ebook management tool that allows you to convert your book files into various formats, making it easier to distribute your work across different platforms.

6 IngramSpark: A print-on-demand and distribution service for both print and ebooks. They offer global distribution and competitive pricing for self-published authors.

7 Canva: A user-friendly graphic design tool that authors can use to create book covers, social media graphics, and promotional materials for their books.

8 Grammarly: An online grammar and spell-checking tool that helps authors polish their writing and catch errors before publishing.

9 Hootsuite or Buffer: Social media management tools that allow authors to schedule and automate their book promotions across various social media platforms.

10 MailerLite or MailChimp: Email marketing services that help authors build and manage their mailing lists, create newsletters, and automate email campaigns to engage with their readers. (There are many more.)

11 Goodreads: An online platform where authors can connect with readers, promote their books, and run giveaways to generate buzz and reviews.

12 KDP Rocket: A software tool specifically designed for authors publishing on Amazon's Kindle Direct Publishing (KDP) platform. It helps with keyword research, category selection, and competitive analysis.

13 Plottr: A visual plotting and outlining tool that helps authors plan and structure their stories, keeping track of characters, timelines, and story arcs.

These tools and platforms can greatly assist authors in the process of writing, formatting, publishing, and promoting their books, whether they choose to self-publish or pursue traditional publishing routes.

Keywords

While every part of the writing, editing, formatting and uploading process is important, the importance of well-targeted keywords might be dismissed by particularly new readers.

So, let me explain.

When authors publish their books on Amazon, they have the opportunity to select up to seven keywords or phrases that best describe their book's content, genre, themes, and target audience. These keywords are crucial in helping readers discover their books among the millions of titles available on the platform.

Here's how the seven keywords help readers find books:

1 Relevant search results: When a reader types a search query into Amazon's search bar, the algorithm scans through the keywords associated with each book to determine which titles are most relevant to the reader's search. Books with keywords that closely match the search query are more likely to appear at the top of the search results.

2 Discoverability in niche markets: By selecting specific, targeted keywords, authors can help their books stand out in niche markets or subgenres. For example, instead of using a broad keyword like "romance," an author might choose "contemporary small-town romance" to attract readers looking for that specific type of story.

3 Increased visibility: Books with well-chosen keywords are more likely to appear in relevant search results, increasing their visibility to potential readers. This can lead to higher click-through rates, more sales, and improved rankings within Amazon's algorithms.

4 Keyword combination searches: Readers often use a combination of keywords to refine their search and find books that match their specific preferences. By selecting a diverse range of keywords that cover different aspects of their book, authors can increase the chances of their title appearing in these combination searches.

5 Improved targeting: Keywords help authors target their ideal readership by focusing on terms and phrases that their target audience is likely to use when searching for books. This can lead to more effective marketing efforts and a higher likelihood of connecting with readers who will enjoy their book.

6 Keyword optimization tools: There are various software tools, such as Publisher Rocket, that help authors research and select the most effective keywords for their books. These tools provide data on keyword popularity, competition, and potential impact on sales and visibility.

7 Long-term benefits: Choosing the right keywords can have long-term benefits for an author's book. As the book gains visibility and positive reviews, it may begin to rank for additional keywords organically, further expanding its reach and discoverability on Amazon.

KEYWORDS ARE USED for all the retailers but are particularly important for visibility on Amazon.

That's because Amazon has become much more than just an online retailer; it has evolved into a powerful search engine specifically for books. When potential readers are looking for their next book to purchase or borrow through Kindle Unlimited, they often turn to Amazon's search bar to find titles that match their interests. This is where the importance of keywords comes into play for authors.

Final words...

You are now well progressed on your writerly journey.

Good luck!

And don't judge your progress by anyone else's.

Every writer's journey is unique, shaped by their individual experiences, goals, and creative vision. Some authors may find success early in their careers, while others may spend years honing their craft before achieving recognition.

There is no single path to success in the writing world, and what works for one author may not work for another.

Some writers thrive in the traditional publishing landscape, while others find fulfilment in self-publishing or hybrid approaches. Some may focus on a specific genre or niche, while others explore a wide range of styles and themes.

Ultimately, each writer must navigate their own path, embracing their unique voice and perspective while learning from the experiences and wisdom of others in the community.

The key is to remain persistent, adaptable, and true to one's artistic vision, recognising that the journey itself is often just as rewarding as the destination.

About the Author

Beverley Eikli is an Australian author of more than 30 Regency romps, and Victorian and Georgian-set romances laced with mystery and intrigue written under the pen name Beverley Oakley.

Under her other pen name—B.G. Nettelton—she writes Africa-set romantic suspense and Women's Fiction.

Born in the African mountain kingdom of Lesotho, Beverley married the handsome Norwegian bush pilot she met in Botswana's beautiful Okavango Delta while managing a safari lodge.

She began her writing career as a journalist, but it was during long aerial survey contracts around the world—often as the sole woman among the crew—that she began to write romance novels, finding in them an escape from the isolation.

Beverley lives just north of Melbourne with the same wonderful husband she whisked away from Botswana thirty years ago, together with their youngest daughter (the oldest lives in Norway), and a gorgeous, dopey Rhodesian Ridgeback who weighs more than she does.

When she's not writing, she runs a bed & breakfast & Farmstay business (called *Wuthering Heights*) in South Australia's beautiful wine growing Clare Valley with her two sisters, and teaches Writing.

You can visit her websites at: www.beverleyoakley.com or www.beverleysbooks.com

Visit Beverley's Shopify store:
www.beverleysbooks.com
Or her website: www.beverleyoakley.com

Join Beverley's reader group on Facebook

Follow Beverley
On Bookbub
On Goodreads
On Facebook

Made in the USA
Monee, IL
04 November 2026

67387312R10075